GIS, ORGANISATIONS AND PEOPLE

A Socio-technical Approach

GIS, Organisations and People
A Socio-technical Approach

D. E. Reeve
University of Huddersfield

and

J. R. Petch
Manchester Metropolitan University

TAYLOR & FRANCIS
ALERE FLAMMAM
Founded 1798

UK   Taylor & Francis Ltd, 1 Gunpowder Square, London, EC4A 3DF
USA Taylor & Francis Inc., 325 Chestnut Street, 8th Floor, Philadelphia, PA 19106

British Library Cataloguing-in-Publication Data

A catalogue record for this book is available from the British Library.
ISBN 0 7484 0653 0 (paperback)

Library of Congress Cataloguing-in-Publication Data are available

Printed and bound by T.J. International Ltd, Padstow, UK
Cover printed by Flexiprint Ltd., Lancing, West Sussex

Contents

GIS for Beginners Series

EDITORS

J. R. Petch
D. E. Reeve

The 'GIS for Beginners' series provides introductory overviews of substantive topics in GIS suitable for study by students at final year undergraduate and postgraduate levels. It is also intended that the books will provide a convenient means for practitioners in industry to gain an overview of subjects of concern to them.

GIS change rapidly and much of the relevant literature is only available in learned journals or advanced monographs. The 'GIS for Beginners' series bridges the gap between the 'advanced' literature and the requirements of students for cheap, accessible introductory texts.

Acknowledgements

There are a number of people we must thank. Our colleagues on the UNIGIS teaching team - Sarah Cornelius, Beverly Heyworth, Ian Heywood, Nigel Trodd and Steve Tomlinson - provide us with a lively academic environment within which to work. UNIGIS students, past and present, have been the source of many of the ideas incorporated here. Steve Pratt, the Huddersfield University cartographer, drew, and re-drew, many of the diagrams. Jean Reeve struggled womanfully to create a coherent text from our scribblings. The Taylor and Francis editorial team have been models of tolerance in the face of repeatedly missed deadlines.

Of course, the greatest debt is owed to our families. If we'd known how much time we would need to steal we probably would never have started.

Our thanks to you all.

Derek Reeve
Jim Petch

Figures

Introduction

'It's not in devising the system,
That fearful dangers lurk.
It's not in devising the system,
But making the system work.
For the working of the system,
Is not in the hands of the great,
But rests in the hands of the poor little clerks,
Like Ahmed and Jose and Kate.'
Anon, quoted in Bell and Wood-Harper, 1992.

'What went wrong? We did our homework. We developed an accurate database. We conscientiously translated their policy statements into maps and composited them as implied by their plan. We thought we had done it all, and we had, from a GIS-centric perspective. What we missed is the GIS wild-card — the human factor.'
Berry, 1996.

ORGANISATIONAL ISSUES ARE IMPORTANT

This book provides an introduction to the human and organisational issues involved in developing a GIS service. It is essential that students and GIS practitioners understand the significance of these issues as a clear consensus has emerged that 'human' issues, as much as technical ones, determine the success of GIS developments. Indeed, Somers (1994) suggests that it is now a 'commonly accepted adage' that organisational issues will determine the long run success of a GIS project. GIS, therefore, can no longer be regarded, or taught, primarily as a technical matter. Building a 'successful' GIS project depends at least as much upon issues such as marshalling political support within the host organisation, clarifying the business objectives which the GIS is expected to achieve, securing sufficient project funding and enlisting the co-operation of end-users, as upon technical issues relating to software, hardware and networking. As the verse above suggests, the success of an Information System ultimately is in the hands of clerks, not computer specialists. The argument in this book, therefore, is that creating a successful GIS within an organisation must be seen as a *socio*-technical, rather than a purely technical, exercise.

GIS is no longer novel. During the last 20 years GIS technology has been introduced into a wide range of organisations. Over 80% of UK local governments have invested in the technology necessary to manipulate digital Ordnance Survey data. Many utility companies have invested in GIS systems to help manage their networks. At GIS conferences there are streams of papers from private sector companies — building societies, insurance companies, retailing companies, banks and advertising agencies — describing their commitments to GIS. In the USA, the 'business geographic' software market grew by 126% from 1990 to 1991 and was expected to continue to grow soon to become the third largest market for GIS products behind utilities and government (Sherwood, 1995). Global sales of GIS software have been growing at an annual rate of 15% and this rate of growth is expected to continue, the market research firm Frost & Sullivan estimating that the total GIS market will expand from $3.85bn in 1996 to $8bn by 2002, an annual compound growth of 13.5% (*Business Computer World*, 1997). Organisations have shown great faith in GIS technologies and large amounts of money are being spent.

Obviously, the organisations concerned are expecting to reap benefits from their investment in GIS in terms of improved corporate performances. Unfortunately, some of them may eventually be disappointed. Reports of early GIS projects suggest, at best, mixed outcomes. Reports in magazines such as *GIS World* and *Mapping Awareness* show that many successful GIS projects have been established and are indeed delivering the expected benefits to their organisations. Human nature, however, gives prominence to success and tends to be rather less forthcoming about failure — it would be a brave employee who used the pages of *Mapping Awareness* to trumpet the deficiencies of his employer's GIS project — but evidence is gradually accumulating of GIS ventures which have been less than successful.

Our own experience, gained from consultancies, and conversations with staff tasked with turning their employers' GIS ambitions into realities often provides a less happy perspective. One hears of corporate GIS projects disintegrating into disparate departmental fragments, of projects which after many years of effort have still to provide worthwhile products, of organisations finding that they have little real need for the advanced GIS capabilities they have purchased, and of funding being reduced as managers become frustrated with the lack of progress. Evaluations by academics sometimes support this equivocal view of GIS success. In the public sector, sustained research by Campbell and Masser (1995) paints a rather gloomy picture of GIS progress in UK local authorities. They report that, despite GIS having been installed for at least two years in each of their 12 case study authorities, only three of them had GIS applications technically operational, and within these authorities only a limited number of the developed applications were deemed of any value by their intended end-users. In two of the case studies, the initial GIS system had already been abandoned and the future of GIS was being reviewed. From their long-term involvement in private sector GIS projects through the GMAP consultancy, Clarke and Clarke (1995) point to the danger that business leaders can quickly become disillusioned with GIS if their GIS projects do not deliver the expected benefits within the expected time period. A Coopers and Lybrand (1988) report suggests that 60% of early GIS projects in North America did not meet the full requirements of their host organisations, and Innes and Simpson (1993) reach similarly equivocal conclusions: 'The twenty-year history of efforts to build large-scale computing systems for planning is ... littered with failed attempts and massive expenditures, often with little to show as a result. Successful institutionalisation of large-scale computing in public agencies has been the exception rather than the rule.' The often-proclaimed benefits of investing in GIS seem to be rather more difficult to achieve in reality than the visionaries (and the vendors) initially imagined and perhaps a degree of disillusionment, even cynicism, is developing.

A GIS 'Joke'

GIS has had more pilots than British Airways. Unfortunately, not so many of them get off the ground.
Anon.

In the early days of GIS, it might well have been the case that flawed technology was the root cause of many frustrations and delays. Early GIS software *was* difficult to use; some GIS packages required training courses measured in weeks; they often had limited functionality, and were prone to crash at critical moments. (Those of us who worked with the early versions of a particular GIS will go to our graves with 'db_vista error' engraved on our hearts). Openshaw et al. (1990) vividly convey the technology 'challenges' with which GIS pioneers had to wrestle:

> 'The system was incomplete when delivered, some of the functionality claimed to be present was missing and some parts of the system were error-prone. It was noted that the system could not perform any of its intended functions and despite several reports stating the problems little or nothing was done to remedy the problems for several months. Worse still, several of the patches when they were made were not successful or made matters even worse.'

GIS software, however, has evolved dramatically during the last 20 years and with each update the quality, capability, reliability and usability of GIS products improve. Commercial pressures have obliged GIS vendors to pay great attention to the usability of their products. Many GIS packages these days present themselves as conventional Windows products — just another icon on the desktop. The hardware platforms and operating systems upon which GIS are now mounted are astonishingly better than they were when GIS packages were first developed. Using GIS software is simply not the fearsome task it was a few years ago.

Another cause of frustration and delay in early GIS projects was the absence of appropriate data. There was a general paucity of digital data, and that which was available was often of poor quality. Having invested in GIS hardware and software, some GIS projects found they had precious little to put in them. Again, however, this situation has changed significantly during the last 20 years. In the UK, for example, the Ordnance Survey has progressively improved the quality and coverage of its digital products and a variety of private sector geodemographics companies have emerged as sellers of digital spatial datasets. The data conversion process is better understood and there are specialist conversion agencies whose business it is efficiently to convert their client's paper materials into digital form. Of course, there will still be many instances when data problems and technical problems will continue to plague GIS projects but the key point is that these sources of difficulty were recognised early in the development of GIS and have been progressively addressed.

As the technical difficulties recede, and data problems lessen, the importance of 'organisational' factors as the cause of failure is thrown into sharper relief. In the mainstream Information Systems industry it has long been recognised that systems fail as much for human as for technical reasons and this reality is increasingly being recognised within the GIS community. Respondents in Campbell and Massers' (1995) local government GIS study, for example, reported that 27.2% of the significant problems that they had experienced were organisationally based, mentioning most frequently poor GIS management structures, staff limitations and lack of support from senior staff. If the success rate of GIS projects is to be improved it is clear that attention needs to be paid to understanding the organisational milieu within which GIS is used.

THE EDUCATIONAL CHALLENGE

Acknowledging the importance of organisational issues in promoting successful GIS projects presents GIS educators and trainers with a challenge. Most GIS programmes are taught within Geography Departments where a majority of staff are unlikely to have had much exposure to the management studies and business computing literatures which provide the necessary context for understanding GIS in business. Many of the staff who now teach GIS come from backgrounds in teaching quantitative techniques and will probably have been attracted to GIS by fascination with the technical challenges it offers. GIS does, so far, appear to have been taught primarily as a technical skill. The emphasis seems to have been upon the spatial analysis and computer science aspects of GIS with, perhaps, some 'organisational issues' lectures thrown in towards the end.

The majority of our students, however, will be hoping to employ their GIS skills not in research or academic environments but rather in public and private sector organisations where GIS is expected to pay its way as a business tool. For such students learning how organisations create Information Strategies will be just as important, possibly more important, as lectures on the latest advances in spatial statistics. Understanding how to create and motivate a project team may well prove more beneficial than an in-depth knowledge of spatial search algorithms. To meet the requirements of such students, there needs to be a shift, at least for some courses, towards teaching GIS within a business management context rather than emphasising its technical content.

This book makes a modest contribution towards the developing literature on the role of GIS within organisations, by providing, hopefully, an accessible, student-level overview of the major issues. Much of the material contained in this book has been drawn from the business computing and information systems development literatures which would not conventionally be found in a book aimed, in part, at Geography students. We believe, however, that GIS students need exposure to this literature if they are to be properly prepared. It is widely accepted that GIS as an area of study lies within the intersection of the 'Computing' and 'Geography' disciplines. More appropriately, however, GIS should be located within the intersection of *three* disciplines — Computing, Geography, and Management Studies.

It certainly is not our intention to write a negative account, but rather to stress that by acknowledging the importance of the organisational issues provoked by GIS, and consciously planning to deal with them, the probability of successful outcomes can be enhanced.

THE STRUCTURE OF THE BOOK

The main themes of this book are quickly summarised:

- We believe that much of relevance for GIS can be learnt by studying developments in the mainstream Information Systems industry. In Chapter One, therefore, we review the shift that has occurred in the mainstream industry away from a virtually single-minded focus on technology towards acknowledging the importance of human factors in computing. We argue that the GIS industry is presently going through a compressed version of this transition.

- Chapter Two provides an overview of current thinking about the role of information systems within organisations. We consider the relationships between business strategies and information systems strategies; the generic benefits which companies can derive from information systems; and the roles which information systems play within companies. We demonstrate that many of the generic benefits and concerns about the use of mainstream information systems with organisations apply equally to the use of GIS.

- In Chapter Three, we tighten the focus of our attention more directly upon GIS, by describing a composite GIS development strategy. Here we consider the User Needs Surveys, pilot studies and cost benefit analysis techniques by which organisations have attempted to determine whether, and which, GIS they might need.

- The conventional GIS methodology described in Chapter Three looks naive and techno-centric compared to the more sophisticated methodologies which have been developed in the mainstream Information Systems industry. In Chapters Four and Five, therefore, we suggest that GIS implementation teams might profitably introduce some of the ideas contained within these modern, mainstream approaches. Chapter Four considers particularly the need to integrate personal and organisational issues into the Information Systems development process and focuses upon two widely used Information Systems Development Methodologies (ISDMs), namely ETHICS and Multiview, both of which emphasise the need for a *socio*-technical view of information systems. Chapter Five considers the impact which emerging computer technologies are having upon the ways information systems can be implemented. Technologies such as Case-tools, object orientation, component-ware, and interoperability are allowing new ISDMs to emerge which promise to allow much more rapid and more flexible delivery. In this chapter we consider particularly Rapid Application Development (RAD), Evolutionary Delivery and the Organic Life Cycle model.

- Chapter Six looks at the range of personal and organisational factors that have been found to be influential in determining the success or failure of information system projects.

- In the final chapter we consider whether there is anything 'special' about GIS which provokes particular problems when they are introduced into organisations or whether, in fact, we would be better to abandon GIS as a separate area of activity, accepting that GIS are in reality simply information systems.

THE ORIGINS OF THE BOOK

This book has been developed from material that originally was written as distance-learning modules provided within the UNIGIS postgraduate GIS diploma programme. The reactions of our students to these modules have been revealing.

Some students had clearly previously conceptualised GIS as a purely technical activity and were rather surprised, and in a few cases dismayed, to find that they were being expected in these modules to consider concepts from the softer, fuzzier, more confusing realms of the social and management sciences. They regarded GIS as computer science and if it didn't involve pounding a keyboard it wasn't computer science. They felt uncomfortable confronting the, to them, woolly issues of organisational politics and personal motivations. Technical problems usually have a solution if enough research and resources are thrown at them, but understanding, and more ambitiously still, altering the ways people behave introduces a different, open-ended, layer of complexity.

The large majority of students, however, have welcomed the 'organisational' modules. Most of our students are already in employment and the issues raised in these modules seem to ring true with their everyday experiences of trying to establish and maintain GIS services within their own organisations. Indeed, the workshops and assignments based upon the modules seem sometimes to have had an almost cathartic effect, allowing students to explain how and why their GIS projects have not developed 'like it said in the book' and to realise they are certainly not alone in this situation.

It is the response of the students upon whom much of this material has been tested that made us believe it warranted a wider audience. Education is, though, a two-way process and we have integrated many ideas and comments from our students into the present text. Over the last six years, our students have taught us well.

THE STYLE OF THE BOOK

This book is an introductory guide. Necessarily we travel swiftly over many topics which should be studied in much greater detail and so for those who wish to study further we have included 'Going Further' sections at the end of each chapter where we cite the references we found most useful and suggest appropriate additional reading. Computer science and management studies are unrivalled for their capacity to generate jargon and so a book that considers a topic within the intersection of these two disciplines must necessarily contain many specialist terms and phrases. To help readers to penetrate the jargon, we provide a glossary at the end of the book: words highlighted in the text ***thus*** are defined in the glossary. At the end of each chapter a set of self-check questions has been provided.

Chapter One:
From Techno-centric to
Socio-technical Computing

In this chapter we consider:

* *the inadequacy of the 'technological imperative'*
* *the swing from 'technology push' to 'demand pull'*
* *the move from techno-centric to socio-technical computing*
* *changing conceptions of host organisations*
* *the changing emphasis from 'computer science' to 'information handling'*

'I experienced an interesting situation last year. Visiting an exhibition, I noticed a man in a smart suit walk onto the stand of a large vendor and ask if he could be shown a GIS. No one from the stand asked the man's particular area of interest or what particular problem did he wish to solve with a GIS. Instead they took him to one of their demo pods and started... within seconds technical jargon was flowing. As I moved off the stand, I suddenly overheard the man say, "Oh I'm with her, I must go".'
Lawrence, in Lawrence and Parsons, 1997.

INTRODUCTION

In this chapter we review the gradual shifts in the computer industry which have lead it away from an almost single-minded focus on technology towards a socio-technical view of computing. We will consider the shift from *'technology push'* to *'demand pull'* in business computing. We will see how this shift is demanding that some, possibly a majority, of computer professionals redefine themselves from being 'computer scientists' to being 'information specialists'.

Our argument will be that the process by which the computing industry in general was weaned away from its obsession with technological issues is being accelerated within the GIS industry: the development of GIS is like watching a video of the history of conventional information systems being replayed at fast forward speed. It took the general computing industry two, perhaps three, decades to arrive at a mature appreciation of the difficulties of integrating computer systems into organisations. It seems that in the GIS world, after only a few years, the gloss associated with our new technological toys is already beginning to wear thin and the effects of some hard questions about the real ability of GIS to help organisations achieve their business goals are beginning to be seen.

THE TECHNOLOGICAL IMPERATIVE AND THE FETISHISM OF THE PRODUCT

Anyone who has stood outside a camera or hi-fi shop drooling over the latest products on show has felt the seductive power of technology. It does not matter that your present video player is perfectly adequate or that the features on the new model are ones Steven Spielberg might use only once a year. You've read the reviews and the same magazine which last year told you how good your present video was now tells you that it's rubbish compared to the new one. The manufacturer's publicity tells you that independent tests prove that the new patented technology built into their product means it dramatically outperforms all the competition. It's got two of those little green digital displays that you don't understand but which look good when they bounce up and down with the sound track. They are offering a discount. You begin to invent reasons why the car does not need a set of new tyres just yet.

Of course, the serious business of buying computer systems is entirely different isn't it? Well no, actually, it often isn't. Perhaps it should be, but it isn't. Purchasing computer systems for organisations usually involves a hidden agenda of personal interests, career ambitions, and inter-group politics, as well as rational, dispassionate evaluations of organisational needs.

For its part, the computer industry relies heavily on the allure of technological advance to sell its products into organisations. The most recent is best. New software is always better than old software. There is no problem that cannot be solved by the latest technological fix. Computer sales people prey on fears that if we do not have the latest and most fully specified product our organisation will be falling behind. They mention that, reluctantly, their company might not be able to continue to support the earlier version, and that anyway everybody is moving to the new product. As a consequence of this, many of us, individuals and organisations, end up vastly over-specifying our computer requirements. We bow to the power of the *'technological imperative'* and are persuaded to buy power we don't need, and functions we don't use (Winfield, 1991).

Are You Over-specified?

If you have access to a GIS, wordprocessor or spreadsheet, take a few minutes to look through the list of functions described in the manual and calculate the percentage of functions you use regularly, the percentage you can remember having used once or twice and the percentage you didn't even know were there. Now divide the purchase price of the software by the percentage of non-used functionality to discover the cost of over-specification.

GIS has been a classic example of the allure of technology. The technological hype surrounding GIS during its initial boom period was astounding. There were GIS gurus, GIS techies, GIS fanzines. Some very large claims were made: 'GIS is a technical innovation as important to the spatial sciences as was the invention of the microscope to the biological sciences' (Abler, 1987); '.... the most significant development in the management of data since the invention of the computer' (Blenheim Online, 1993); '... the biggest step forward in the handling of geographic information since the invention of the map' (Department of the Environment, 1987). GIS vendors' advertisements focus on gleaming machinery attended to by attractive smartly dressed, sun-tanned people. Such images bear no resemblance to any GIS labs we have ever seen. Where were the empty coffee cups, the dog-eared manuals, the bits of masking tape you can never quite get off the digitiser, the thumbprints on the screen? They do, however, successfully convey the message that GIS products were technologically advanced and highly desirable items. Anyone who visits the exhibition hall of one of the major GIS conferences will know that the emphasis remains almost entirely upon technology. Vendors compete by demonstrating the technological marvels of their products: our GIS is better than theirs because we have object-orientation; our new version is faster than theirs; the new release has 30 new functions. As visitors we look cool but really we'd like to knock the demonstrators out of their chairs and have a go ourselves. We focus upon the spectacular graphics which the demos provide and we are not far removed here from the consumer drooling over electronic delights in shop windows — with the significant exception that with a carefully crafted report and a little internal politicking we might just get our bosses to buy our new toys for us. Huxhold and Levinson (1995) observe 'for most people buying GIS technology is the most exciting phase of a project — for many people it is akin to buying a car'.

Whether we are buying electronic goodies for ourselves or computer systems for our organisations, we are not engaged in a rational process. Marx got it right when he wrote of the *'fetishism of the product'*, by which he meant that capitalism teaches us to desire products for themselves — for the status which possession will convey — rather than for what they will actually do for us. Purchase provides gratification: possession confers status. Possession becomes an end in itself. Constant (1994) similarly suggests that we buy the image of the product rather than the product itself. The computer industry is no different to any other industry in that it wants to create within us as individuals, and as organisations, a desire to purchase products. Where the computer industry has an edge over many other industries is the association of new technology with modernity and progress. The technological imperative leads us to believe that the adoption of new technology is inevitable and the sooner we adopt the new products the greater will be our advantage and status: 'Much of what may be termed the folklore surrounding technology evokes images of progressive successful environments while those that cast doubt on the appropriateness of such views are regarded as backwards and even killjoys' (Campbell and Masser, 1995).

A GIS 'Joke'

There once was a super GIS salesman that travelled the world with a great 'it can do everything' GIS demo (but the real stuff was vapourware). He sold it to lonely GISers and made lots of money. One day while dashing through an airport on his way to clinch another mega deal he dropped dead of a heart attack.

At the gates of Heaven he was judged. He had lived a borderline life and was given the option of Heaven or Hell. He could look into the doors of each and choose. As he opened the door to Heaven, wonderful harp music played, he saw people floating on clouds and all was bright and white.

Next he opened the doors to Hell and saw people drinking beer and dancing to rock and roll music. Everyone was partying. It was just like his first year at college.

When he met with his Maker again, he said: 'Heaven is great and wonderful, but the other is more my style'. 'Think carefully', he was told, but the other was his wish.

As the doors of Hell opened for him, the intense heat hit him and he was pulled in. He stood before the Devil and saw pain and sorrow everywhere. He shouted at the Devil: 'Where is the party and beer?'

The Devil laughed: 'That was the demo, this is the real thing'.
Bruce Baiklie, e-mail, GIS-L, 1993.

FROM 'TECHNOLOGY PUSH' TO 'DEMAND PULL'

> 'If I define a successful system as one that is developed *on time* and *within budget*, it is *reliable* and *maintainable, meets its goals* and *satisfies its users*, how many of you would say that your organisation builds successful systems? I've asked this question of hundreds of people at all levels of data processing, and the overwhelming response is silence ... the vast majority of us have worked on systems that do not meet these criteria for success — in other words, on systems that in some way can be classified as failures.'
> Block, 1983.

As the computer industry has proved, it is possible to sell a lot of products upon the promise and image of technological advance, but, eventually, there always comes a day of reckoning. One's family concludes that they can't tell the difference in picture quality between the new and the old video, and what do those green graphs mean, anyway, Dad? At work, accountants and managers begin to ask just where are all these benefits which were so confidently predicted from the latest upgrade? Eventually, the focus of attention shifts from 'Gee you mean it really does this?' to 'Did we actually *need* it to do this?'

By the early 1980s computer professionals such as Block were concerned about the poor track record of the systems they were delivering. In too many cases information systems were not delivering the benefits that were promised to the organisations which paid for them. An early North American study by Mowshowitz (1976), for example, reported that only 20% of the information systems considered were regarded as having been successful, while 40% had achieved a marginal gain and 40% were failures. In 1988 Eason, reflecting on a history littered with disappointments, concluded that introducing information technology had proved to be a high-risk undertaking, where complete failure is not uncommon, and marginal impact common place. Eason compares the difficulty of introducing information technology into an organisation to that of transplanting an organ into a human being. If all goes well and the transplant is accepted by the host great benefits can be achieved, but there is a great risk of rejection and failure. Sometimes 'success' is achieved only by taking actions so drastic that they severely weaken other parts of the organisation.

Recent highly publicised failures such as the London Ambulance Control system (Barker, 1998) (which incidentally included GIS software) and the Taurus project for the London Stock Exchange are evidence that the problems of catastrophic information systems failures are still with us. Indeed, a survey of IT projects in the USA reported by Legg (1996) suggests that success rates have improved only a little since Mowshowitz's 1976 study. Legg writes that 16% of projects in 1995 were completed in time and on budget, 31% were cancelled before completion and the rest were 'challenged projects', i.e. were completed but with limited functionality. To an outsider the pages of the weekly computer trade press continue to provide fascinating, if sometimes gruesome, evidence of an industry struggling to deliver. There are stories of systems failing, products failing to meet their specifications, suppliers being sued by clients, clients switching suppliers halfway through projects and projects being unceremoniously ditched. The bald, unhappy truth is that information systems continue very often to fail to deliver the benefits expected of them.

When information systems do fail, the immediate, almost instinctive reaction is to look for technical explanations. What went wrong is that the software couldn't cope, network infrastructures or protocols were inadequate or the response times of the system were poor. We regard information systems as technical projects, therefore we look for technical reasons for failure. Furthermore, the recommended remedy will probably involve more of the same — more, newer and better technology will solve the problem.

For many years, however, some computer professionals have concluded that the major reasons for the lack of success of their systems lies not so much in any technical limitations, but rather in their neglect of the human and organisational aspects of computing. Systems have been delivered which ill-fitted the organisations into which they were to be introduced. Organisations have been expected to bend to accommodate the new technologies, rather than the other way round. Systems have been imposed on organisations because the technology made them possible, rather than because there was a genuine demand for them. In short, Information Systems have been driven by technology-push rather than by demand-pull.

The response to the recognition of these problems has been a gradual shift in emphasis in the way computer scientists regard themselves and their products. Increasingly a *socio-technical* viewpoint has been adopted (Figure 1.1), with the definition of an 'information system' being expanded to include not only the hardware and software but also the people involved. Increasingly there has been a movement towards a 'demand led' view of computing — information systems being regarded as worthwhile only if they are meeting genuine user needs.

TECHNO-CENTRIC COMPUTING	SOCIO-TECHNICAL COMPUTING
• focus on technology • technology push • because it's possible • others are doing it • hierarchic • specified by technologists	• people and technology • demand pull • because it's needed • WE need it • democratic • specified by users

Figure 1.1 From techno-centric to socio-technical computing.

It has taken mainstream computing something like 30 years to develop its current, mature view about the relationships between technology, organisations, people and business. It seems that the GIS industry will go through this cycle in about 10 years. Until very recently the momentum of the GIS industry was fuelled by the excitement generated by a rapidly advancing technology, but there are now signs that this technological impetus is beginning to run out of steam, and questions are being asked about the ability of GIS to deliver business advantage. Certainly within the GIS literature, one can see a growing impatience with the prevailing techno-centric view of GIS. People seem to be wondering whether the technology has been oversold. Reid (1992) effectively summarises the position:

> 'There has been something of a backlash recently in the world of GIS, with many questioning the value of the technology for their day to day business needs. GIS is often now seen as high on cost and future potential but very low on short-term deliverable benefits. Indeed, at a recent GIS software user conference the key-note speaker, an independent consultant, did his hosts few favours by suggesting that GIS had yet to prove it had any real benefits for the world of business and commerce.'

Consider also the comment from the UK's Local Government Management Board (LGMB, 1995):

> 'GIS projects have all too often been led by technology or by the attractions of owning the most advanced computer. However, like any information system, GIS is only as good as the data which is put into it and should be driven by the needs of the business.'

People have begun to ask questions about the business efficiencies gained by companies that have adopted GIS. How many of the pilot studies written up in *Mapping Awareness* have actually been carried through to successful long-term business systems? What are the human problems of introducing digital mapping into organisations where staff have made their careers out of traditional cartography? How many GIS 'failures' or 'disasters' are lurking as yet unacknowledged within pioneer organisations? Definitive evidence of the success, or otherwise, of early GIS projects is difficult to obtain but the little evidence that there is suggests that the success rate of GIS projects may be no higher than that for conventional computer projects. Buchanan (1993) suggests that GIS generally have the characteristics which make them fall into the category of 'risky projects'. Coopers and Lybrand (1988) reported that in the USA 60% of GIS pilot projects which they examined did not meet the full requirements of their host organisations. Campbell and Masser's (1994) research into success rates of GIS projects in British local authority projects paints a similarly down-beat picture. Of one of their case studies, they write that '£1 million had been spent on the development of a GIS with virtually no return'. Unless the GIS industry can quickly begin to prove the utility of GIS in business situations the healthy scepticism which seems to be growing among potential users might turn to outright disillusion.

Perhaps we are witnessing the beginning of the transformation of GIS from a technology-pushed towards a more mature, user demand-pulled industry in which the emphasis is not on technology but on the successful integration of GIS into organisations. Perhaps a socio-technical conception of GIS is emerging in which it is recognised that even if it is assumed that GIS software succeeds on a technical level, its adoption will ultimately depend on how well implementation strategies address organisational barriers (Cullis, 1994).

EVOLVING CONCEPTIONS OF THE 'ORGANISATION'

Recognising that organisational issues can be critical factors in determining the success of Information Systems, has meant that Information Scientists have had to develop much richer conceptions of what 'organisations' are, how they behave, and particularly how they are likely to respond to the introduction of new information technologies. As Checkland and Scholes (1990) put it:

> 'In the 1960s the adoption of the standard assumption from management science that organisations could be treated as if they were instrumentalities, goal-seeking machines, seemed not unreasonable. But in the 1980s such an assumption seemed increasingly dubious. Why not treat organisations as if they were not goal-seeking machines but discourses, cultures, political battle-grounds, quasi-families, or communications and task networks?'

Campbell and Masser (1995) provide a convenient tripartite summary of the evolution of attitudes towards the host organisations into which information systems are introduced:

Technological Determinism: Here the host organisation is assumed to provide an unproblematic, indeed almost unconsidered, environment into which new systems can be introduced. The inherent superiority of new technology means that the organisation will inevitably adopt it. Any problems within the organisation are symptomatic of old-fashioned thinking or Ludditism.

Managerial Rationalism: Here it is recognised that introducing new technology will cause some problems of adjustment within the organisation. The conception of an organisation, however, is still that of a rational, almost machine-like, structure which is amenable to logical adjustment. Problems caused by the introduction of new technology, therefore, can be accommodated by logical restructuring of organisational procedures. The currently fashionable Business Process Reengineering methodology is consistent with this view of organisational behaviour.

Social Interactionism: Here, as suggested in the quotation from Checkland and Scholes, organisations are viewed as very complex, social structures which cannot be expected to behave rationally. Organisations are viewed as being composed of groups of individuals each with their own motivations and ambitions. In such a conception, the adoption of new technology, no matter how impressive, is by no means assured. Whether an information system is a success will depend upon a complex interaction of, often informal, political and social forces within the host organisation. If the people inside an organisation cannot be persuaded to adopt a new system, the system is little more than very expensive junk.

Increasingly the *Information Systems Development Methodologies (ISDMs)* adopted by mainstream information scientists to help them introduce information systems into organisations have been designed to accommodate the human realities implied by the 'social interactionist' conception of organisations. We review some of these methodologies in Chapters Four and Five. These methodologies regard the process of introducing information systems as much as a process of reconciling the concerns and aspirations of employees within organisations as of designing the technicalities of the systems.

Although in mainstream computing ISDMs that incorporate a broadly socio-technical conception are now commonplace, in GIS the emphasis appears to have remained so far resolutely focused upon technical issues. As Campbell and Masser (1995) put it:

> 'The most notable feature of the social interactionist perspective in relation to GIS is its virtual absence from the literature in this field, either in terms of an explanatory framework or as a basis for implementation. ... the majority of studies related to GIS have concentrated on refining technical know-how, largely divorced from the environments in which the technology in which it is expected to operate.'

Only in the last three or four years has a stream of literature begun to emerge which takes as its focus the organisational environments within which GIS technologies are used, rather than focusing upon the details of the technologies themselves. The central argument of this book, however, is that, just as the mainstream Information Systems scientists found before us, GIS specialists need to become sensitised to the realities of organisational behaviour.

COMPUTER SCIENTISTS OR INFORMATION SPECIALISTS?

The increasing recognition of the socio-technical nature of Information Systems, coupled with the increasing maturity of hardware and software, has also required mainstream computer specialists to reconsider their own 'self image' and roles within organisations.

> 'We have too many computer scientists. To me computer science is a transitional phenomenon. It will become a branch of mathematics and a branch of electronics. The subject of the future is information systems. But there's a real problem here. The staff (in universities and business) are still technologists. I'm not sure they have the commitment to run with the change. You must be committed to information systems, not information technology, if the transition is going to work.'
> Angell, 1993.

One reason why the techno-centric view of computing persisted for so long is probably to do with the personalities and training of the people attracted into Computer 'Science'. Computer Science has traditionally seen itself as a 'real' science, i.e. a hard, quantitatively oriented science closely related to mathematics, logic and electronics. The titles on the traditional computer syllabus — 'Systems *Analysis*' or 'Software *Engineering*' — convey the self-image of the discipline. Inevitably Computer Science tends to attract students who are intrigued by technological puzzles.

To a considerable degree this emphasis on the technical issues was justified in the early days because getting early computers to work *was* a major technological challenge. When computers consisted of vast banks of electronics components which broke down so frequently that computer vendors had to provide permanent on-site engineers, many of the problems of computer science were technological. The complexities of early operating systems and low level programming did put a premium upon mathematics and logic skills.

In recent years, however, there has necessarily been some reassessment of the self-image and training of computer professionals. The increased reliability and simplicity of use of systems has reduced the need for the traditional computer 'boffin'. There continues to be a need for a minority of true computer scientists to work in the research laboratory developing new systems, but the majority of students going through computer courses today are likely to spend most of their working lives providing information services, using packages written elsewhere. The primary role of many computer professionals today is to liaise with users, rather than to hide in the computer room. Kuiper (1992), for example, divides computer professionals into two groups, 'computer scientists' and 'information handlers' and estimates that the ratio between the two to be about one to twenty. The attitudes, and indeed personalities, needed by the information handlers differ from those of the pure computer 'boffin'. (Flynn (1993) suggests star signs may even differentiate them — boffins should be introvert Virgo or Cancer whereas information handlers will be extrovert Aries or Leo!) The information handlers still need to understand the technology, but they also need the skills more often associated with management professionals. The days when the 'techno-nerds' could insulate themselves behind the glass walls of the computer rooms are going, possibly already gone.

Figure 1.2 summarises the changing role of the computer specialists within organisations. In the early days of data processing, users would often have to try to explain their requirements directly to computer programmers. Soon, however, the new profession of systems analysts emerged with skills to enable them to extract from users their requirements and translate these requirements into technical specifications appropriate for programmers to work with. Latterly there has been some further refinement of roles with a distinction being made between the business analyst, who liaises with users in order to build a business specification for a system and the technical analyst who carries out the formal analysis. Increasingly the user has been buffered from the technicalities of computing by information specialists whose job it is to translate business needs into computer specifications. Figure 1.2d, however, indicates the dramatic impact of personal computing upon the role of the computer specialists. Many users now, for better or worse, expect to be able to develop their own applications, with computer specialists being regarded as experienced advisors or facilitators, rather than as experts leading the process.

Again, there are lessons for the GIS industry to be learned from the mainstream industry. Already only a very small percentage of the people who would regard themselves as GIS specialists are actually involved in writing new GIS software. Most of us spend our days using GIS packages to set up databases, and customising interfaces, to provide services for clients and colleagues. Thus, in terms of Figure 1.2, most GIS specialists are presently fulfilling systems analysts' roles — we help users to define their needs for geographical information and then establish and operate the required geographical information systems for them. Most current GIS packages are designed to be used by specialists but with each new release of software, the packages become easier to use. The era of 'shrink-wrapped' GIS packages which end-users will expect to use directly will soon be upon us, if indeed it is not already here.

In terms of GIS education this means that we too should be stressing the human skills involved in using GIS successfully. For most GIS students the abilities to understand the roles of GIS in organisations and to be able to help users formulate their requirements will be more valuable than will be an ability to code ever more efficient 'C' algorithms. Perhaps GIS should not only be taught in Geography and Computer Science departments but also increasingly in Management Studies Schools?

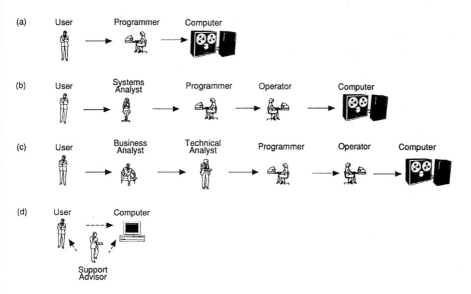

Figure 1.2 The changing role of the computer specialist. (Source: modified from Avison and Fitzgerald, 1988, *Information Systems Development Methodologies, Techniques and Tools*. Used with permission.)

CONCLUSION

We have looked briefly at the changing attitudes towards information systems that have occurred within business over the last 30 years. In the early years, information systems were seen primarily as technological products and were sold on the basis of their technological merits. The disenchantment caused by the high failure rate of the early systems, however, forced a reconsideration of the role of information systems in business. In essence there was a shift from 'technology push' to 'demand pull' and from a techno-centric to a socio-technical view of computing. Information Systems professionals have developed increasingly sophisticated conceptions of how organisations behave and thus increasingly sophisticated ISDMs. Computer staff within organisations are fulfilling the role of 'information handlers' rather than traditional 'computer scientists'.

The shorter history of the GIS industry is unfolding like a fast replay of the development of the mainstream industry. After only a few years of technological hype, some hard questions are now being asked about the ability of GIS to deliver benefits to the organisations that pay them. There is strong merit in the argument that GIS specialists should adopt a socio-technical conception of GIS.

WHAT YOU HAVE LEARNT IN THIS CHAPTER

* *   *early developments in data processing were largely technology driven there is a relatively high rate of failure among information systems projects*
* *   *the computer industry is now aware that computing is best regarded as a 'socio-technical' enterprise*
* *   *there is a need to acknowledge the realities of organisational behaviour in order to improve the success rates of information systems projects*
* *   *the skills needed by computer professionals have changed as the balance of demand has moved from 'computer science' to 'information provision'*
* *   *the GIS industry seems now to be moving from a technological to a user orientation*
* *   *some doubts are being expressed about the reality of the benefits often claimed for GIS*

GOING FURTHER

An indication of the slowly gathering recognition of the importance of 'human issues' as a determining factor in GIS projects is the recent increase in the number of books which focus upon issues relating to GIS and organisations. Six years ago, when the first version of the distance learning modules from which this text has developed, there were no texts that focused upon this topic. Now there are a number of texts that would repay reading by anyone who wishes to study these issues in greater depth. These include:

CAMPBELL, H. and MASSER, I., 1995, *GIS and Organisations: How Effective Are GIS in Practice?* (London: Taylor and Francis).

GRIMSHAW, D., 1994, *Bringing Geographical Information Systems into Business*, (Harlow: Longman).

HUXHOLD, W.E. and LEVINSOHN, A.G., 1995, *Managing Geographical Information System Projects*, (New York: Oxford University Press).

MEDYCKYJ-SCOTT, J. and HEARNSHAW, H. M. (Eds.), 1993, *Human Factors in Geographical Information Systems*, (London: Belhaven Press).

OBERMEYER, N.J. and PINTO, J.K., 1994, *Managing Geographic Information Systems*, (New York: Guilford Press).

There is a considerable mainstream literature upon the need to develop socio-technical approaches to information systems development. Attention is drawn to some of these texts in the 'Going Further' sections of later chapters. Two early texts which give an excellent overview of the socio-technical approach and which it is appropriate to mention here are:

EASON, K.D., 1988, *Information Technology and Organisational Change*, (London: Taylor and Francis).

WINFIELD, I., 1991, *Organisations and Information Technology: Systems, Power and Design*, (Oxford:Blackwell Scientific Publications).

Chapter Two:
Information Systems in Organisations

In this chapter we look at strategic planning for information systems in organisations. We consider:

* *the generic benefits of investing in information systems*
* *the roles which information systems fill within organisations*
* *the importance of developing coherent information strategies*
* *the difference between an Information Systems (IS) and an Information Technology (IT) strategy*
* *some tools for strategic thinking*

'A big problem for management is determining whether a request for additional computer funding is truly justified by a business case or merely a desire for new technology toys.'
Kuiper, 1992.

'A computer specialist who has been taught the technical skill of developing software but has not learned about business operations is not unlike a teenager with an American Express card but no concept of a family budget.'
Kuiper, 1992.

'The joy of being left undisturbed to solve an intricate puzzle on the computer can become a consuming passion. Any real value that the puzzle may return to the company is only a secondary consideration.'
Kuiper, 1992.

'Anyone who studied and specialises in programming shouldn't be let anywhere near ... management without first being given ego and personality transplants in addition to retraining.'
Ahmed, 1993.

INTRODUCTION

In many organisations GIS packages operate as specialist tools within particular departments or projects, requiring a relatively small outlay and of concern only to small groups of staff. Beyond their immediate work groups, such 'project level'

GIS systems usually will have little impact upon the broader functioning of their host organisations.

In many organisations, however, GIS are now regarded as being of major, *corporate* significance. In local authorities, for example, 'land and property' databases can be viewed as one of the core databases required for efficient management (on a par, for example, with their personnel and finance databases). Similarly, the Facilities Management GIS systems being developed by the utilities companies will, if successful, be central to these industries' business operations. These corporate level GIS systems provoke permanent, major changes in the way organisations work, alter people's working lives, and demand large amounts of resources to complete. In computer-speak, large GIS systems can be **mission critical**: getting the corporate GIS wrong would have very serious consequences indeed for the overall performance of the company.

It must be expected, therefore, that when a proposal for a corporate GIS development is put forward it will be subjected to the same strategic, high level scrutiny as would be any major new Information System proposal. Consequently, it is essential that proponents of GIS have a vision of how strategic IS proposals are developed and evaluated within modern organisations in order to be able to position their proposals appropriately. In this chapter we provide an overview of the benefits which modern organisations expect will be achieved by spending on information technologies and present some ideas about how modern organisations develop Information Systems strategies.

THE TWO DOMAINS

A first point to make is that when GIS specialists propose a new system, they are, in fact, stepping onto a battlefield across which corporate politics and tensions rage. Organisations have traditionally had great difficulty in defining the proper roles that IT should play within their structures and in many companies long-standing tensions exist between mainstream management and IT specialists. From the business managers' viewpoint, the information technologists sometimes seem a tribe apart, with separate training programmes and different career structures. Like a mediaeval priesthood, IT specialists are caricatured as having hallowed sanctuaries, mysterious rites and private languages. Like a priesthood, they are suspected of using their specialist knowledge as a means of retaining influence within their organisations. It is argued that IT specialists by inclination and training are often unable fully to appreciate the business contexts within which they are employed: 'Information technology has always been seen as the "ivory tower" department within the typical company. Many users perceive the IT manager to be a grouchy person in a white coat with a severe attitude problem. Moreover, the department has generally not been known for listening to the demands of users' (Bradbury, 1995).

Business managers, who often come from non-technical backgrounds, sometimes feel themselves ill prepared to evaluate the potential of information technology. If their IT manager tells them the company needs a new system, who are they to argue with the expert? Conversely, some managers may have recently gone too far the other way and have begun to encroach too far upon the IT specialist's territory. The spread of PCs and of 'end-user' computing has caused many people to assume that they 'know about computing' and has caused some to challenge the expertise of IT staff; not recognising the significant differences between personal computing and corporate systems. There is a, possibly, apocryphal story of a 'sales manager who spent the weekend with his son setting up a database using Microsoft Access and then refused to accept from his IT manager that it would take six months to construct an enterprise-wide sales and marketing database. The world is full of people who, upon constructing an Excel spreadsheet begin to think they're Babbage himself' (Bradbury, 1995). A little knowledge...

Some business managers have become particularly resentful of what are seen as the unnecessary constraints imposed by centralised computing services. They wish to purchase their own IT fixes to their departmental problems and become frustrated when computer services start talking about corporate standards, maintenance contracts and support costs.

The Two 'Domains'

Parker et al. (1988) use the term 'domain' to distinguish between the business and the IT cultures within a company:

Business Domain	**Technology Domain**
Business units that use corporate information services	Staff and equipment to provide information services to business units
Purchases services from technology domain	Sells IS/IT services to business domain
IS/IT is a COST	**IS/IT generates INCOME**

Parker argued as early as 1988 that most of the technology problems within the technology domain were solved, the biggest problem for the technology domain already being to prove the value of its services to the business domain: 'We know how to write programs and purchase hardware. We know how to create databases. What we typically don't know is how to apply the technology best to bring benefits to the business domain.'

Against this troubled background, it would be naive for GIS enthusiasts to assume that major spending will be sanctioned on their pet projects simply because it is obvious to themselves that GIS is 'a good thing' for their companies. Business managers need to be persuaded that GIS can deliver benefits that are consistent with the business objectives of the organisation. Mainstream IT staff, jealous of their own turf, may also need to be shown that the GIS will be compatible with the Information Systems structures already established within the organisation. GIS proponents thus need to understand how best to pilot IT proposals through the corporate minefields.

In an attempt to lessen the tensions generated by corporate IT investments, a considerable amount has been written about how organisations should plan their

Information Technology developments and how IT specialists and business managers should define their roles with respect to each other. In this chapter we provide a brief review of some aspects of what is now considered best practice in this thorny area.

GENERIC BENEFITS OF INFORMATION SYSTEMS

At a most general level, there are three ways in which organisations have conventionally been envisaged as being likely to benefit from investments in Information Systems technology. These are:

- Efficiency benefits
- Effectiveness benefits
- Competitive advantage benefits

These three generic benefits are not watertight categories and often a single system will offer some elements of all three. Silk (1991), however, suggests that in mainstream computing there has been a chronological progression from efficiency, through effectiveness, to competitive advantage (Figure 2.1) and it is convenient to consider each separately. A similar tripartite classification of benefits can also be used with reference to GIS, although the timeframe for GIS developments is much shorter. We defer consideration of the 'Enterprise Computing' and the 'Virtual Organisation' until later in this chapter.

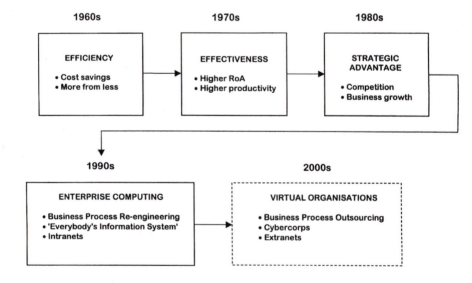

Figure 2.1 Chronology of generic benefits of IS/IT. (Source: modified from Silk, 1991, *Planning IT: Creating an Information Strategy*, Oxford: Butterworth-Heinemann, Figure 3.3. Used with permission.)

Efficiency Benefits

Efficiency benefits accrue where the introduction of a computer system allows a given level of output to be achieved with a reduced resource cost, the objective here being to reduce operating costs.

The efficiency saving which is most often, and often controversially, associated with the introduction of information systems is a reduction in labour requirements. Large numbers of clerks, bookkeepers, bank-tellers and manual workers performing repetitive tasks have been displaced by computer systems. In mainstream computing, however, introducing computer systems has also led to other resources being used more efficiently. For example, where a computer system is being used to replace previously bulky paper systems, space savings can result. Also computer control of production processes in industry, and of heating systems in buildings, leads to increased efficiencies in the use of material and energy. It is clear that GIS offers very great scope for labour efficiency gains in map-based processes. The contrast between the labour intensity of traditional paper-based map systems — time wasted in retrieving maps, drafting and duplication bottlenecks, difficulties of map based analysis, etc. — against the efficiency of digital mapping is obvious.

The 'CFTM' Factor

'Workflow analysts studying conventional manual filing systems discover inefficient storage and retrieval practices so often they now have a dismissive acronym for it CFTF: "can't find the file".'
Wainwright, 1996.

Anyone who has worked in an organisation with a large manual mapping system will appreciate the very real frustrations caused by the equivalent CFTM factor — 'Can't Find The Map'. The development site you are interested in always seems to be split across the corners of four map sheets and two of these will have been 'borrowed' by someone else.

Although GIS enthusiasts tend to focus on the advanced processing that the technology offers, for many employees the most basic benefit offered by GIS — an accessible, current, edge-free map base — is probably the most valued.

Inviting organisations to go back to using typewriters to produce documents and then to use Banda machines to reproduce them, or to re-employ the armies of clerks who used to calculate payrolls, would be greeted with howls of derision. Yet prior to the adoption of GIS, the way organisations processed spatial data was the equivalent of using a typewriter (or even a quill pen!) to produce text and logarithm tables to carry out calculations. What the wordprocessor and spreadsheet were to typing pools and accounts offices in the 1970s and 1980s, GIS is to mapping offices in the 1990s. The rows of drawing boards and light-tables which once were common in mapping authorities, utility companies and local government departments are disappearing. The prima facie case for GIS offering substantial, long run, labour efficiencies to organisations which process large volumes of mapped data is unassailable.

In addition to efficiencies in the use of labour, GIS can also generate efficiencies in the use of other scarce resources. For example, a justification which has been put forward for corporate GIS in local authorities is the reduction in floorspace permitted by moving from paper to digital mapping. In a major authority there will be a considerable floorspace occupied by mapchests and maptanks. Moving to digital mapping releases expensive floorspace for other uses.

The Ordnance Survey of Ireland: Efficiency Benefits

In 1992 the Ordnance Survey of Ireland installed a Sysdeco 'Tellus' system to provide digital storage and distribution of their maps. In consequence of this new software:

— A saving in excess of IR£1m has been achieved by releasing space for other needs. New buildings would have to have been built if this space were not available.
— The overhead of maintaining duplicate masters for security reasons is being eliminated by replacing them with low cost, space saving digital storage.
Source: Diploma student.

Effectiveness Benefits

Effectiveness benefits are achieved when computer systems facilitate higher levels of achievement from investment in other resources. Rather than reducing costs, the emphasis here is upon improving productivity. IT is used as a facilitator to allow companies more closely to achieve their goals, to take better decisions and to provide a better service to their customers.

In general computing there are, of course, numerous examples of IT enhancing staff productivity. Issuing sales staff with laptop systems to provide on-site quotations enhances their productivity and provides a better service to customers. Video conferencing minimises time lost through business-related travel and thus increases the organisation's effective use of executive time. At the most basic level, wordprocessing packages allow secretaries to produce more letters, electronic mail speeds communications and spreadsheets ease financial calculations.

Although the prima facie case for GIS offering staff cost (efficiency) savings is clear, the logical consequence of this — staff reductions — is often politically unacceptable, and so the efficiency benefits of GIS are often left implicit in published appraisals. More often the benefits of GIS are presented more positively as *effectiveness* gains — the same number of staff will be able to produce more and better work. Cartographers will produce more maps, policy makers will be able to work with better data, planners will make more informed decisions and clear more applications, and engineers will produce more network diagrams. In a world where staff are expensive, and employers are reluctant to increase payrolls, introducing GIS as a tool to increase the effectiveness of existing staff has a powerful appeal.

Providing a Better Service

'Officials in Phoenix after successfully completing a GIS recently bragged that the technology has reduced from five days to five minutes the time it takes to verify ownership and get other property records to respond to citizens' complaints and questions about maintenance, zoning and other environmental issues. They receive 22,000 such requests annually.'
Authes, 1993.

'It is now possible to go into any meeting armed not only with a general map of the site in question, but with one which contains all the constraint information which will influence the meeting. This is something which could not have been contemplated before the acquisition of a GIS.'
Source: Diploma student.

In addition to increasing the effectiveness of staff, GIS systems can enhance the effectiveness with which organisations use other resources. An early name for GIS — Automated Mapping/Facilities Management (AM/FM) systems — indicates that one effectiveness benefit would be from better management of plant. The basic claim is that knowing where facilities are located and being able rapidly to recall and update details of their status will improve the efficiency with which an organisation can manage its facilities. This argument underpins much of the rationale of the utilities' investment in GIS. The command and control systems introduced by organisations such as the police, ambulance and vehicle rescue services can also be seen as attempts by these organisations to improve the effectiveness of their mobile facilities. Maintaining an emergency patrol vehicle is expensive. If a GIS system can be used to increase the number of repairs completed per shift, and in so doing reduce the waiting times of clients, it can contribute significantly to the effectiveness of the service. Leyden (1998), for example, reports that in order to meet tougher Government standards for emergency response times, the Sussex Ambulance service are installing a GIS/GPS-based control system, which trials indicate will reduce the time taken to dispatch an ambulance to an incident from an average 90 seconds using the current voice system to 12 seconds using the new system.

'Coopers and Lybrand recently concluded that Westminster City Council would start to get a return on investment in GIS after four years and ultimately could expect a £250,000 per annum benefit for GIS. The major benefit would be in staff time savings, but this would be likely to result in increased productivity and improvements in the quality of services [i.e. effectiveness] rather than in absolute savings in staff numbers [i.e. efficiency].

There would also be benefits from identifying "missing properties" during the creation of a master property index. Reducing the number of properties which escape the council's notice will significantly increase its local tax revenues.'
Source: Planning Newsletter, July 1993.

Competitive Advantage Benefits

Competitive advantage benefits are achieved where information systems are used to gain a competitive edge over rival organisations. There are many ways in which this can be achieved.

Information technology can be used to distinguish a product from its rivals. For example, a much-promoted attraction of tele-banking is that, unlike normal banks, tele-banks never close. (Apparently, for many people the major practical advantage of this is that they can organise their financial affairs late on Saturday nights when there is nothing left worth watching on the television (Silk, 1992).) Similarly the Direct Line insurance company distinguished itself from its rivals and achieved a startling initial growth rate by promoting the use of direct communications. (The Direct Line example, however, neatly exemplifies a reality of the 'technological imperative' which is that competitive advantages gained from technology are usually transitory. Noting the success of Direct Line, competitor insurance companies quickly responded by establishing their own direct contact insurance operations. Firms that compete via technology are on a treadmill. They must continually invest and reinvest and only the most recent will do: 'Being a pioneer in competitive advantage is like running too fast in the 10,000 metres. You're out in front for a few glorious moments, then the pack swallows you up' (Grindley, 1993).) Some firms gain competitive advantage by using information technology to be able to respond more quickly to market changes. The high-street retailer W. H. Smith, for example, invested £3 million pounds in software that allows it to analyse the vast amounts of data that flow into its central *data warehouse* from the tills in its stores. This software allows W. H. Smith to search for consumer trends. During November 1993 the software spotted a decline in the sales of computer games in the middle of an expected increase. The company cancelled a £500,000 order and switched its Christmas promotions to other products, two to three weeks before rival retailers became aware of the trend.

With regard to GIS there are numerous examples of systems being used in the private sector by firms to gain competitive advantage. The banks, building societies and major retailers have invested in GIS as tools to measure their market penetration, guide their advertising campaigns, advise upon site locations, target potential customers, and monitor the performance of their rivals. A good example of these types of use are provided by the GMAP consultancy's GIS systems which integrate interaction modelling techniques with GIS and which have been used by firms such as Toyota, Ford and W.H. Smith to help them plan their distribution chains (Birkin et al., 1996; Grimshaw and Scholten, 1997).

Just as there was a progression in the mainstream Information Systems industry from a 1960s concern with gaining efficiency benefits to current concerns to gain competitive advantage benefits, a similar progression can be seen with GIS projects, though over a shorter timespan (Grimshaw, 1994). The early adopters of GIS were primarily organisations which had traditionally been large users of conventional maps — environmental agencies, local governments, the utilities and national mapping agencies. These early GIS systems were designed to replace the manual mapping systems and in so doing to deliver efficiency and effectiveness savings.

Competitive Advantage: The MACE Shopper 2000 Project

The convenience shopping market in the UK is becoming ever more competitive as petrol station forecourt convenience sales increase and the large supermarket chains move back into urban locations, as the potential for new out-of-town superstore developments declines. The Mace group of conventional convenience stores is using GIS techniques as a tool to help it protect its position in this competitive environment. They commissioned the CACI geodemographic company to create a classification of their stores based upon the purchasing characteristics of their market areas ('Rural', 'Retirement and Telegraph', 'Tabloid and Ciggies', etc.) and are now using this classification to fine-tune the product range, promotion, opening hours, etc. of each of the twenty-two Mace store types to match that type's likely customer profile. Convenience stores depend vitally upon local customers. Using GIS analysis, Mace can now segment their strategies to match local characteristics and thus gain competitive advantage.

Source: Thurman, 1996.

In recent years, however, the fastest expanding sector of GIS spending has been into previously non-geographically based organisations such as banks, building societies, retailers, insurance companies and advertising agencies (Sherwood, 1995). In such organisations there usually will not be a pre-existing, manual map system to replace and so conventional map-based 'efficiency' cost savings are not generally available. On the contrary, as GIS is a new, additional activity within many private sector companies, GIS might actually be regarded as adding initially to the total cost of their operations (although in practice the costs of private sector 'marketing' GIS projects are usually much smaller than those incurred by the large 'operational' GIS systems established by the traditional map user organisations). The motivation behind many private sector investments in GIS, therefore, is the search for competitive advantage. Companies believe that spending on GIS will gain them information which will allow them to 'gain an edge' over their competitors.

ROLES OF INFORMATION SYSTEMS WITHIN ORGANISATIONS

Another way of considering the benefits which can accrue from investing in Information Systems is to identify the roles which they fulfil, and the levels at which they are used, within and between organisations (Huxhold, 1991).

Here we meet the 'triangle' diagrams so beloved by writers on organisational issues. Regardless of their particular spheres of activity most organisations develop broadly similar organisational structures. At a most general level, this 'typical' organisational structure is usually represented as a triangular structure (Figure 2.2). At the base of the triangle is the operational layer where production processes take place. (Usually the organisational triangle is discussed in terms of manufacturing organisations, but clearly the triangle can easily be translated to consider service and public organisations, where 'products' are the services provided to customers and residents.) The middle layer consists of managers, researchers and administrators whose tasks include monitoring the performance of the operational layer, researching the external environment of the organisation, and preparing policy options for the top layer. The top layer of the organisation, here labelled the executive layer, consists of the relatively small group of decision makers who determine the strategic direction of the organisation.

The triangle model has found its way into textbooks in a very wide range of subjects including sociology, geography, economics and business and management. Its relevance here is that each layer has distinctive information requirements and hence demands distinctive kinds of Information Systems. It is thus possible to classify Information Systems according to the roles they play within the different levels of the triangular structure.

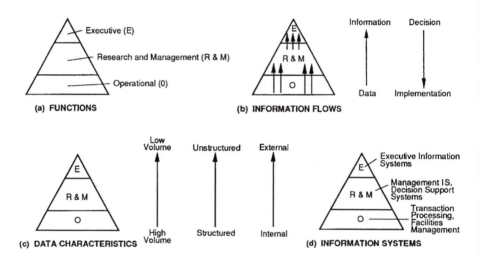

Figure 2.2 The triangular structure of organisations.

Operational Level

The data manipulated at the operational level tends to be voluminous, rapidly changing, but of low variety. Typically in a private sector organisation, one thinks of order processing and invoicing operations. In the public sector, examples include processing planning applications, tax forms and benefits claims. The requirements for successful information processing are capacity, accuracy, and speed. Any decisions that need to be taken are likely to be 'programmed' in the sense that there will be standard rules and procedures to be followed. It was at this level that the earliest impacts of computerisation were felt with the introduction of basic *Transaction Processing Systems (TPS)* in the 1960s.

Many of the GIS systems introduced by mapping agencies, utilities and local governments are the spatial equivalents of traditional TPS. The emphasis here has been to use GIS to make basic (spatial) operational activities such as map production, plans processing, and facilities management more efficient (Figure

2.3). Usually these systems have replaced previous manual map processing systems. Such 'information processing' GIS require the ability to handle large volumes of spatial and *attribute data* but usually only require limited spatial analysis capabilities; most of the enquiries submitted to them being of a routine and pre-programmed nature. Indeed, as Campbell and Masser's (1995) research confirms, the most frequent use of GIS in UK local government at present seems overwhelmingly to be the most basic GIS function of all — producing site plans from Ordnance Survey data.

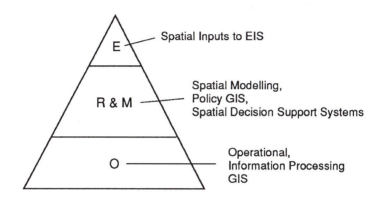

GEOGRAPHICAL INFORMATION SYSTEMS

Figure 2.3 GIS within the organisational triangle.

Management Level

The information required at the management level should come from two sources. Managers need to know how the operational level is performing and so there should be a major *internal* flow of information from the operational level to the management level. The second flow of information into the management layer should be from *external* sources since it is at the middle level that officers have to monitor the business environment. Into this management level should flow the streams of government statistics, market research, and anecdotal information which organisations require to be aware of threats and opportunities in their environments. The functions of the middle layer include analysing information coming from internal and external sources, comparing performance against targets set by the executive layer, and taking any necessary corrective actions. The middle layer also refines data from internal and external sources to produce position statements and policy options for the executive layer. The characteristics of the data used by the middle layer are that they are more varied but less voluminous than those at the operational level. They may also be more ephemeral than at the bottom layer. Officers at this level will often work intensively on a project for a short period but then move on to other responsibilities. The decisions taken at this level are semi-structured and non-routine.

To help managers in this middle level of the organisational triangle, a class of information systems generally referred to as ***Decision Support Systems (DSS)*** has developed. These systems help managers take decisions by providing tools and models with which to analyse policy relevant information. The range of the systems which fall under the heading of DSS is very wide and might include general purpose statistical and modelling packages, through to bespoke programs modelling industry-specific processes. They are typically highly interactive systems making use of graphs and diagrams to help analysts visualise the consequences of decisions. A good example comes from the financial markets where dealers are judged by the number and quality of the decisions they make, and banks of computer screens are provided by their employers to maximise their effectiveness. Fortunes can be made and lost by fractional delays in the delivery of information to dealing rooms and so it is critical that the Decision Support Systems provided are highly optimised. Some modern DSS include aspects of artificial intelligence in that they not only present the information to the decision maker but also advise upon feasible courses of action.

There are many examples of GIS systems being used as policy and research tools in the middle layers of organisations. Indeed, the term ***Spatial Decision Support Systems (SDSS)*** has emerged to describe this method of employing GIS technology (Densham, 1991). In the private sector, there are many examples of GIS systems being used to assist in spatially-based decision making — where to locate new outlets; which outlets to rationalise; market area analysis; which areas to target for marketing purposes, etc. Grimshaw (1994) describes examples of GIS being used to support decision making in retailing, financial services, motor vehicle dealerships and fast food franchises. In the public sector, many local governments and health authorities are now using GIS as a means of enhancing their spatial policy making.

Whereas 'operational level' GIS systems tend to be 'heavy weight' systems requiring an ability to store and process large volumes of data, although, perhaps, with only rudimentary spatial analysis capabilities, GIS systems which are used as Decision Support Systems within the middle layer of the organisational triangle tend to have different characteristics. Small, relatively inexpensive, desk-top systems, but with good analytical and presentational capabilities, will often suffice.

Problems in the Middle Layer

Officers in the middle levels of local governments often in the past had an almost impossible task, in that they were required to develop detailed analyses and policy recommendations from woefully inadequate information bases. Figure 2.4 contrasts what should happen (a) with what often actually happened (b).

Information that should come from the operational layer was often trapped in data files which could not be interrogated conveniently for policy purposes and so the flow of information from below was weak. Also in the UK at least the availability of external information has been limited by doubts about the currency, accuracy, range, and expense of both government and private data sources. The very fine level of spatial detail at which they are often required to work further exacerbates the problems of spatial policy makers. Officers, for example, can be asked to justify 'action areas' of various kinds which may cover only a few streets.

(a) In Theory

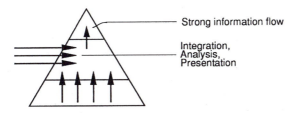

(b) In Reality

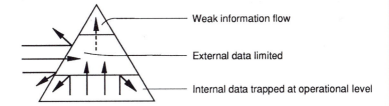

Figure 2.4 Information problems in the middle layer.

Increasingly GIS has been seen as a means of ameliorating the information difficulties of policy and research officers. GIS is being used as a means by which policy information can be abstracted from operational data by aggregating these data onto a spatial basis. GIS also provides the tools to integrate internal datasets with external data sources, and thus to hold the combined 'policy indicator' databases in a form convenient for analysis and presentation.
Reeve and Wheeler, 1991.

Executive Level

Decisions at the executive level tend to be 'strategic' in that they are long range, non-routine, complex, unstructured and critical to the organisation's survival. Ideally the information needs of executives should be met primarily by highly refined summaries of internal and external conditions prepared by middle layer officers. Executive officers should not be concerned with detail. Indeed, management consultants often find that major problems arise when executives forget their roles and become too involved in analysis and management.

Because of the nature of the information involved, Information Systems were relatively slow to penetrate this level of organisations. During the 1980s, however, *Executive Information Systems* (*EIS*) began to emerge. EIS are intended to provide executives with the information they need, when they need it, and in whatever form they find most convenient. An analogy with the instruments on a car's dashboard can be used to explain the purpose of an EIS. On the executive's screen should be the key indicators which show her that the organisation is going in the right direction, at the right speed, and that nothing is about to blow up. Alter (1992) summarises the types of information which might be included on the executive's dashboard system under the following headings: 'comfort information', i.e. information such as last month's sales figures which executives will rarely act upon but which makes them feel comfortable; 'warnings', e.g. that a project is badly behind schedule; 'key indicators', i.e. measurements of important organisational performance; and key external information, e.g. 'gossip', anecdotal information, facsimiles of newspaper cuttings, which are of relevance to current issues.

It is doubtful whether conventional, heavy duty, GIS should ever penetrate a Chief Executive's office. Such GIS systems are designed primarily to be used by specialists and as such are unsuited to direct use by executives. Clearly, however, there is scope for the results of GIS-based analyses to be available on the Chief's desk and the GIS vendors are progressively making their products more easily used by non-specialists. Development toolkits, such as 'GeoTools', ESRI's 'MapObjects' and Intergraph's 'Geomedia' products are now available which allow software developers to integrate mapping functions easily into business software and the GIS vendors are also rapidly developing technologies which allow mapped data to be displayed and interrogated via the now effectively, ubiquitous, Web browsers (see below). As GIS vendors put greater emphasis upon the usability of their systems it is reasonable to expect digital mapping and spatial analysis in some form to penetrate ever higher into an organisation's hierarchy. Executives will probably not realise that they are 'doing GIS' but they will increasingly find mapped information delivers to their screens. As Lawrence and Parsons (1997) put it, this will be 'GIS in disguise'.

Across the Levels: 'Enterprise Computing'

So far our presentation has implied that the information needs of each layer will be met by separate computer packages — the operational layer will have its Transaction Processing Systems, the middle layer has a variety of DSS, and the executives might have their EIS. In reality this has indeed often been the case: systems have been bought at different times, from different suppliers to meet different needs. This heterogeneity of systems, however, can considerably weaken the flows of information within and between levels in organisations. There are 'islands of automation' but between the systems there are discontinuities which inhibit the flow of data. Particular problems can arise where staff are expected to make good such discontinuities. Staff inevitably come to resent the burdens of having to re-key data, produce special reports, etc. simply to feed the information requirements of staff at another level in the organisation.

In an ideal world, the information systems within the organisational triangle would be seamlessly integrated — the EIS transparently feeding from the DSS which in turn would automatically abstract its internal data requirements from the TPS. Indeed it can be envisaged that the information requirements of all three layers could ultimately be met by a single software system, which could be customised to present an appropriate interface and set of functions to each layer of employees.

 Some commentators have begun to suggest that recent advances in technology are allowing this ideal to come close to reality in some organisations. The acronym 'EIS' has been hijacked to mean not 'Executive Information System' but rather *'Enterprise Information System'* or even *'Everyone's Information System'* in order to emphasise the pervasive, all-embracing, influence which information technology will have within modern organisations. Rather than having 'islands of information', the vision is that in modern, information-based organisations every employee should have convenient, desk-top, access to the information they need.

For example, among a number of emerging technologies, the development of *Intranets* has attracted particular attention as a mechanism for disseminating information efficiently within companies. Intranets are the private, internal equivalents of the Internet, the idea being that the tools that have been developed for the Internet — World Wide Web browsers, Email, search engines and the like — can be used to provide a very convenient and cost-effective method for delivering corporate information to employees. Conventionally, organisations wishing to spread computer-based information across their operations might have needed to establish *client/server* networks which typically require considerable systems expertise to set-up and maintain. With Intranet solutions things are easier. Information can be formatted into Web documents and posted onto the Intranet by office staff rather than computer experts; customised front-ends are not necessary; and portability across operating systems is not an issue — one simply installs the appropriate browser. Furthermore, at present at least, much of the necessary software is freely available. It has been suggested that the Internet, and corporate Intranets, may become so important that conventional PC machines will soon be replaced by *Network Computers* (*NCs*), these being 'cut-down' machines which will simply pull data and applications ('Applets') from networks as they are required.

As Intra (and Inter) net technologies develop, map-based information products will appear ever more readily upon the desk-tops of employees. With conventional GIS software, companies have had to estimate the number of GIS 'seats' they need to purchase and this seat restriction subsequently has limited access to the GIS software within their organisations. GIS software can be expensive and firms have not wanted to install it onto machines where it may be used only occasionally. Linking GIS into their Intranet systems, however, will mean that companies will be able greatly to expand access to map-based information without having to purchase large numbers of full GIS systems. In future, instead of employees having to have GIS packages permanently stored on their PCs or Workstations, they will be able temporarily to pull down an appropriate GIS applet from their Intranet server. Furthermore, if the specialist GIS tool they may require for a particular task is not available on their own Intranet, they may be able to 'rent' it for a limited period from an external GIS vendor's site via the Internet.

GIS and the WWW

'Overnight, the Web transformed everyone's perception of an integrated application. Now web-enabled GIS can put geographic data into the corporate mainstream by making it simple to access information through a browser...Almost anyone can use a browser. Now instead of populating an organisation with $10,000 GIS seats, you achieve zero-cost distribution.'
Boyle (Intergraph Vice President), quoted in Wilson, 1997.

The expected advantage of establishing 'enterprise computing' will be that employees will be able to use readily accessible, current and consistent information on their desk-tops to take better and more prompt decisions: the 'IQ of the organisation' will be raised.

Information Systems between Organisations: Towards Virtual Organisations?

A limitation of the triangle model is that it focuses upon the growth of information systems *within* individual organisations. Increasingly, information systems also have a role to play in improving efficiency, effectiveness, and competitive advantage *between* organisations (Figure 2.5).

At the Operational level, firms were quick to see the advantages of linking their internal TPS systems into larger inter-firm electronic data processing networks. The illogicality of firms having to print paper orders from their internal TPS systems, sending the orders through the post to suppliers, who then had to re-key the orders into their own TPS systems is obvious. *Electronic Data Interchange (EDI),* in which computers in different organisations exchange operational data, has been an established aspect of business activity for some time.

1. OPERATIONAL LEVEL

2. MANAGEMENT LEVEL

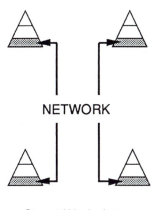

NETWORK

e.g. Street Work data

Order and Supply data

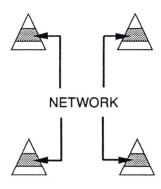

NETWORK

e.g. Joint GIS Projects

'Host' Networks

Figure 2.5 Inter-organisational information systems.

With regard to GIS, there is clearly considerable scope for developing inter-organisational electronic transfer of maps and geo-referenced information. Indeed, in the UK there have been discussions for a number of years between the utilities and local governments about the feasibility and desirability of electronic transmission of street work details (Hartley et al., 1992). More recently there has been a proposal that organisations involved in the UK property industry would benefit considerably from an electronic national land and property register and a pilot project is running in Bristol to investigate the feasibility of developing a National Land Information Service (NLIS) which would create a 'virtual database' by drawing from data held by the Ordnance Survey, the Valuation Office, Land Registry and local authorities (Smith and Goodwin, 1996). Progress in developing inter-organisational sharing of geographical data in the past has been hampered considerably by the variety of proprietary formats in which such data have been traditionally held and, to some degree, by difficulties in transmitting large data volumes via slow networks. (Newton et al., 1990) reported that at the beginning of the 1990s, the predominant mechanisms for transferring spatial data by organisations involved in geoprocessing to external contacts remained tape and diskette.) The move towards *interoperable* GIS software and open data standards, developing national and international geo-data standards, and the ever increasing bandwidths of communications, however, are rapidly lessening the technological difficulties associated with sharing geographical data between organisations. During the next few years there will certainly be a considerable growth in inter-organisational transfers of operational spatial data, although, in the context of this book, it should be observed that sharing geo-data between organisations raises a host of inter-organisational 'human' issues which are only now beginning to be explored. If organisational issues are often barriers to successful GIS implementations *within* individual organisations, what problems will there be when attempting to share geographical data *between* organisations? When data is shared between two or more organisations, whose definitions of the meanings of the data will be adopted? Who will have the right to update the data? Who is responsible for quality and who for errors? If company A takes action on mapped data supplied by company B, who is legally responsible? Who will guard confidentiality (Somorjay and Yeoman, 1996)? One UK utility company currently distributes copies of its GIS data to other utilities in its area via read-only CD-ROMs, presumably in order to ensure that it knows exactly what has been supplied.

At the management and research level of the organisational triangle there are also advantages to be gained by inter-organisational networking. Good early examples of the benefits to be gained are the 'HOST' systems which were established in Manchester and Kirklees local authorities as tools to promote local economic development. In the HOST concept, a local authority sponsors the establishment of a local network through which local businesses and community groups can contact each other and obtain access to a wide range of commercial databases. The expectation is that enhancing interaction between local firms and increasing the availability of relevant market information will promote local growth. Initially, the HOST development was text-based, but as bandwidths increase, one can envisage a HOST-like network which offers as part of its services a 'Community GIS' for use as a shared resource by local organisations. The explosion of Internet availability and the emergence of technologies to deliver map-based information over the Internet is, of course, hastening the realisation of such developments.

Looking to the future, the increasing interconnectedness of organisations has led many commentators to believe that the distinction between activities which are 'inside' and those which are 'outside' any organisation will become increasingly blurred. Some organisations are already experimenting with **Extranets** which allow trusted suppliers, partners and customers to participate in dataflows which previously would have been retained in-house. Here **Business Process Outsourcing** is a key concept. Whereas in the past communications difficulties have meant that firms needed to retain many functions in-house, improvements in informatics mean that organisations increasingly can now allocate support functions to outside agencies in order to concentrate on core activities. Already many organisations have outsourced their mainstream information systems requirements to specialist 'facilities management' companies and clearly there is scope for outsourcing GIS. Rather than each major organisation within a region — local authority, health service, emergency services, etc. — maintaining its own digital mapbase and GIS specialists, a model for the future might be to establish an external specialist GIS service provider. Indeed, in some British metropolitan counties there have been for some years 'information' units, for example the 'West Midlands Joint Information' and 'Greater Manchester Research' units, which offer GIS services to their constituent district councils.

A continuation of the trend begun with extranets may well lead to be emergence of *virtual organisations.* Such organisations would appear to their customers to be 'real' organisations but would in reality exist only on the Internet, being in fact a number of separately owned, and possibly physically widely dispersed, activities co-ordinated by communications technology. Grimshaw (1996) explores the roles which GIS might play within such virtual structures, arguing that there will still be a need for spatial awareness in order to co-ordinate the physical realities of material movement and delivery to customers.

DEVELOPING AN INFORMATION STRATEGY

So far we have considered the broad categories of benefits which might lead an organisation to consider investing in particular types of Information Systems. We turn now to consider how organisations develop an overall view of their Information Systems requirements. How does an organisation ensure that maximum benefits are obtained from its investment in IT? How does it decide its IT priorities? Which systems will fit best into its overall information infrastructure? How to blend the inputs of business managers and IT specialists to obtain a coherent IT program? How to know when to use in-house staff and when to recognise that consultant expertise is needed? In summary, how should an organisation establish an *Information Strategy*? Here we provide a skeleton of the most relevant ideas.

Corporate Strategy

Although the quotes at the top of this chapter suggest that fascination with technical puzzles might cause some IT specialists to lose track of this basic truth, it should be obvious that an organisation's Information Strategy is ultimately subservient to its Corporate Strategy: the Information Strategy must be 'mission driven', in the sense that any computer purchase can only be justified if it contributes to the organisation achieving its primary business objectives. Managers need *information,* not computer systems: 'The focus of most business managers has shifted from being dazzled by the wonders of technology to an expectation that it should either support their business or be ignored' (Massey, 1997).

Just to make sure, we'll put it in big letters:

INFORMATION SYSTEMS ARE THE SERVANTS OF ORGANISATIONS
(not the other way round)

The starting point for a discussion of Information Strategies, therefore, must be a preliminary note about *Corporate Strategies*. A Corporate Strategy document is a keystone statement that should be prepared by senior executives to define the purposes and objectives of their organisation. In establishing the corporate strategy, executives and directors will need to weigh a broad spectrum of internal and external influences, opportunities and constraints. External factors might include, for example, pressures from competitors, market conditions, new legislative requirements, and new opportunities provided by technological change, including IS/IT developments. Internal influences may include reviewing existing product and marketing policies, auditing facilities and resources, and reviewing personnel and skills policies. Once the keystone Corporate Strategy has been decided, sub-strategies relating to functional areas such as Marketing, Production and Information Systems can be generated.

In many companies it has become fashionable for corporate strategies to be formalised into a hierarchy of elements (Figure 2.6).

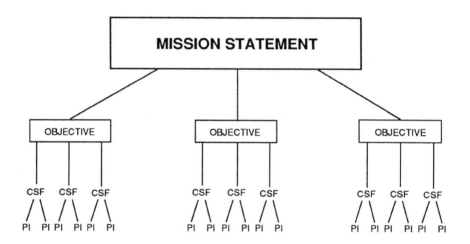

Figure 2.6 Corporate Strategy — Mission Statement, Objectives, CSFs, PIs.

Mission Statement: This should be a brief statement, sometimes a single sentence, which encapsulates the fundamental, long-term aims of the organisation. The *mission statement* should not be concerned with detailed, short-term issues rather it should clarify the main purpose of the organisation and provide a backcloth against which more detailed objectives can be worked out: 'a beacon that guides subsequent planning and development rather than states specific objectives' (Huxhold and Levinson, 1995).

Objectives: From the mission statement it should be possible to derive a series of more specific *objectives*. Each objective should be a specific target for the organisation that should be achievable within a stated timespan.

Critical Success Factors: For each object there will be a number of events and influences 'where things must go right for the objective to be achieved' (Edwards et al., 1991). These are the *Critical Success Factors (CSFs)*. Ideally CSFs should be specific and measurable. It should be possible to devise

Performance Indicators (PIs) to measure progress towards achieving CSFs. Analysing CSFs and PIs can guide managers towards appropriate corporate Information Strategies, as they indicate the essential information requirement which managers need to measure the performance of their parts of their enterprises (Robson, 1994). Key PIs should ideally be on the dashboard of the Chief Executive's Executive Information System.

In some companies the hierarchy of objectives, CSFs and PIs is matched explicitly to the organisational structures within the company — each director being responsible for the achievement of a particular objective or group of objectives, and each group, indeed each individual employee, having their own formally agreed CSFs and PIs which can then be used as a basis for annual performance reviews.

Information Strategy

Once an organisation has in place a clear view of its mission in the form of an agreed Corporate Strategy, subsidiary strategy documents can be formed within each of the organisation's functional areas to translate the Corporate Strategy into reality. Traditionally, areas such as production and marketing would have been considered. Increasingly, however, one of these sub-strategies will now be a formal consideration of the organisation's information processing requirements. (It has been estimated that by 1991 between 60–70% of UK companies had developed formal Information Strategies (Silk, 1991).)

Definition

'Strategic IS/IT planning is taken to mean planning for the effective long term management and optimal impact of information, Information Systems (IS) and information technology (IT), incorporating all forms of manual systems, computers and telecommunications. It also includes organisational aspects of the management of IS/IT throughout the business.'
Ward et al., 1990.

Silk emphasises that the process of determining an Information Strategy must start with broad considerations of business needs and only latterly should the focus turn to specific information technologies. He suggests a three-stage process:

Review the broad objectives of the business, and consider the role of 'information' as a resource in helping to achieve these objectives. Whereas in previous eras, collecting and disseminating information within a company might have been regarded as a cost burden and a distraction from the main purposes of the company — think of the pejorative term 'pen-pushers' used to describe the clerks who used to perform these tasks — in modern organisations information is regarded as a key resource on the premise that those companies which can best marshal their information will have a considerable competitive advantage. The term *Information Resources Management (IRM)* is used to describe the philosophy of active management of information as a key resource. The starting point of developing an IS strategy, therefore, should be to acknowledge the importance of information management, to reflect at a broad level upon the role of information within the organisation and to develop a commitment to active information resource management.

Research the manner in which individuals and groups within the organisation use information. What information do they require? Where does it come from and in what formats is it transmitted?

Identify the Information Technologies which might be introduced to improve the efficiency of the storage and flows of this information?

Notice that the first two stages listed are about information requirements and people, not technology, and this leads us to the next important point which is that, in current practice, Information Technology (hardware, software and technical expertise) is very much the servant of information requirements. And again...

INFORMATION TECHNOLOGY IS THE SERVANT OF INFORMATION REQUIREMENTS

(and not the other way round)

The distinction between Information Systems requirements (IS) and Information Technology requirements (IT) is often formalised by there being two separate and unequal documents prepared (Figure 2.7):

Information Systems (IS) Strategy: The IS Strategy is a statement of the information needs of a company, i.e. a *demand* statement. Its production will involve an input from IT professionals but it must strongly reflect the requirement for Information Systems as perceived by the business managers. It should make clear what Information Systems are required in order to service the *Business Strategy* and clarify who is responsible for achieving what is required. In practice, consultants are often engaged to provide an impartial review of an organisation's IS requirements. In addition to computer-based information systems, the Information Strategy documents should also encompass an analysis of the roles of manual information systems in fulfilling the organisation's information needs.

Information Technology (IT) Strategy: This is a statement of how the organisation intends to meet the needs for computer-based information systems as expressed in its IS Strategy. The IT Strategy is a *supply* document, detailing the software and hardware requirement, and is primarily the province of the IT specialists.

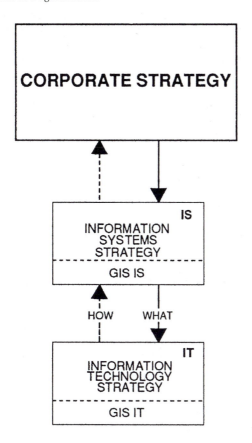

Figure 2.7 IS and IT strategies. (Source: modified from Edwards et al., 1991, *The Essence of Information Systems*, Prentice Hall: London, Figure 4.1. Used with permission.)

The manner in which the IS/IT planning process is performed will vary considerably from organisation to organisation. In some organisations formal methodologies are adopted. For example, IBM's BSP (Business Systems Planning) provides a methodology which explicitly takes as its starting point an organisation's Business Strategy and works through a number of stages to determine the 'Information Architecture' the organisation requires. In other companies less formal methods of IS/IT planning may be used. Detailed accounts of IS/IT strategic planning methods are provided in the texts listed in the 'Going Further' section at the end of this chapter. Here, we will make just three further observations:

The 'IS' Vacuum: Until recently the distinction between IS and IT was not fully understood in many companies, and in particular IS strategies were weakly developed. In the absence of clear IS statements, IT specialists have been often obliged to fill the gap trying to anticipate what their business colleagues would require. In consequence, IT departments have often been the scapegoats for systems failures which in truth were caused more by lack of management guidance than by imagined technical inadequacies. If business managers do not specify clearly what is required, IT staff can hardly be blamed for not meeting the specifications.

The Narrowing Gap/Hybrid Managers: It has been convenient to portray the IT staff and business managers as warring camps, but in reality the gap between the two is probably lessening as both sides learn more about the others' concerns. Much has been made in recent years of the need for 'hybrid' managers who would have both technical and business training and thus be able to bridge the gap between the two camps. The tension between the two camps can be highlighted by asking who should best be appointed to head an organisation's IT department — an IT expert, who understands the technicalities, or a mainstream business manager who might be better able to direct the IT staff towards the organisation's business goals? Of course the best answer is someone who understands both computing *and* business, but such hybrid managers are still in short supply.

Top–Down Approach: Strategic IS/IT planning is an emphatically top–down approach. The Strategic approach attempts to maximise the benefit from spending on IT by imposing a coherent view from the top. Whether it is realistic to expect such top–down planning to be effective is a moot point. The dangers of a '5 year plan' approach in an area as rapidly changing as IT are self evident.

How not to do it:

Council invites experts to rethink IT plan

[A district] council is bracing itself for an IT overhaul following a decision to call in external consultants. The decision was made after the collapse of the council's computer strategy working party, set up three years ago with a remit to determine future IT policy. According to a document prepared for consultancies bidding for the business, the working party had no priorities and 'gave approval for the projects to those who shouted loudest and longest'.

The council has no IT department. Instead, a team of 20 IT staff reports to the finance department. The arrangement has led to friction with other departments. A briefing document says: 'The IT function is viewed by departments other than finance as an isolated function which adds costs to its service budget which are unjustified.'
Martin, 1993.

Relatively little has been written, so far, about the need to integrate GIS into corporate strategic IS/IT procedures but the implications for GIS are clear. Strategic IS/IT planning is a means of disciplining corporate IS/IT expenditures and eliminating maverick, ad hoc projects. If corporate GIS proposals are to succeed they must be integrated into the mainstream corporate IS planning procedures and must expect to go through the same hoops and pass the same scrutinies as other major computer projects. A report into the barriers delaying the adoption of GIS within UK central government departments, points to the dangers inherent in GIS failing to become part of mainstream IS planning:

> 'A common problem is that GIS is often treated as a business application in its own right, and is set to one side as a self-contained system, whereas, for best effect, it needs to be established as an integral part of an organisation's IS infrastructure and to be applied according to business needs. Until GIS is correctly identified in IS Strategy plans, it will remain a poor relation to other IS facilities even within organisations in which geographic information is dominant.'
> Cambridge Computer Consultants and Dan Rickman Associates, 1993.

Clearly, the advocates of GIS within organisations need to understand how to project their systems into mainstream IS planning debates.

TOOLS FOR STRATEGIC THINKING

Finally in this chapter, we will look briefly at a number of devices which have been developed to help managers develop their strategic thinking. These devices are not formal, numerical techniques. Indeed, in reality they often represent little more than invitations to think long and clearly about a problem. They are, however, widely used in IS/IT planning, and GIS managers will benefit from knowing something of them as devices to help them develop their own proposals.

Project Mission Statements

The hierarchy of mission statements, objectives, critical success factors and Performance Indicators which has been described above at a corporate level can also profitably be applied in miniature to give clarity of purpose to individual IT projects. Somers (1994) explicitly recommends the application of a mission statement philosophy for major GIS projects:

> 'A clear understanding concerning the role of GIS in the organisation will provide the answers to many problems down stream in the implementation process. "Vision" or "Mission" are terms commonly applied to this statement of role. Whatever it is called the concept involves a clear articulation of the planned role of GIS in the organisation and agreement among all participants regarding this role.'

Unless a GIS project has an agreed mission, clear objectives and defined Performance Indicators, how can anyone in an organisation know whether GIS is succeeding for them?

SWOT Analysis

SWOT is a technique which is often used to help organisations arrive at a realistic definition of their mission statements. 'SWOT' stands for **S**trengths, **W**eaknesses, **O**pportunities, and **T**hreats Analysis — these being the broad headings which managers developing a strategy or considering an investment should review. Strengths and Weaknesses relate primarily to internal and current issues. What is the company good at? What are the weaknesses in the present operation? Opportunities and Threats are primarily to do with external and future issues. What are our competitors doing? Will emerging technology strengthen or weaken our markets and products?

Within the SWOT approach there is scope for considerable variation. At its simplest SWOT may do no more than provide the headings for a 'back of the envelope' consideration of the 'pros and cons' of a proposal. Alternatively, each element of SWOT might be the subject of formal and detailed scrutiny. Either way the acronym serves as a valuable aide-mémoire of the strategic issues decision makers should consider. The outcome of a *SWOT analysis* should be a vision of what the company, or project, wants to achieve. Strategies provide the maps by which the vision is to be arrived at, and CSFs are the milestones along the way (Figure 2.8).

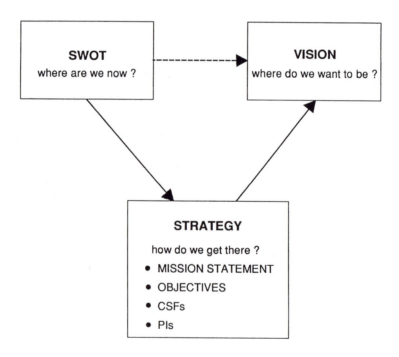

Figure 2.8 SWOT and the strategic planning process. (Source: modified from Silk, 1991, *Planning IT: Creating an Information Strategy*, Butterworth-Heinemann: Oxford, Figure 4.5. Used with permission.)

Boston Matrix

This technique is conventionally used for assessing the prospects of particular products. It is named after the Boston Consultancy Group who first popularised the approach. Basically the approach proposes that any product can be assessed along two axes. The first axis relates to current market share. The second axis relates to the potential for market growth. The two axes allow projects to be placed in one of four categories — 'dogs', 'wild cats', 'stars' and 'cash cows' (Figure 2.9a).

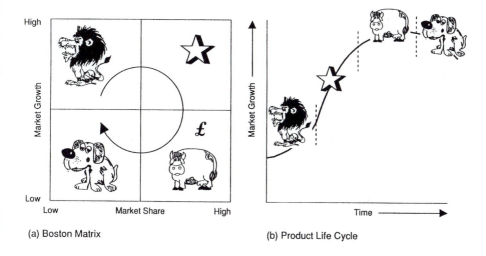

(a) Boston Matrix   (b) Product Life Cycle

Figure 2.9 The Boston matrix.

The matrix can be related to the product life-cycle concept, with any product which survives a complete life-cycle from launch to maturity being likely to go roughly through the matrix sequence from wild cat to dog (Figure 2.9b). A 'wild cat' is a product that presently has a small share of a rapidly increasing market and so has growth potential, assuming an adequate development strategy can be devised. Such products are likely to have additional resources, including IS/IT resources, allocated to them. Hopefully, wild cats turn into 'stars' which are products that have a large share of a growing market. 'Cash cows' are mature products for which the market is no longer growing but which continue to earn worthwhile returns without requiring additional investment. 'Dogs' are products with low market shares in a low growth potential market. Disinvestment is likely to be the appropriate strategy for such products (Edwards et al., 1991).

In this form, the ***Boston matrix*** can be useful to IS/IT planners in a number of ways. Imagine, for example, a GIS unit which operates as a cost centre within an organisation, selling its products to other chapters. Such a centre can be regarded as having a portfolio of information products each of which could be subject to matrix analysis, with the wise GIS manager obviously hoping to ensure a bias towards 'star' and 'cash cow' products. Perhaps providing a basic digital mapping service to client departments would provide such a GIS unit with a 'cash cow' product, which the manager could then milk for resources with which to develop emerging GIS products.

Value Chain Analysis

Every product or service goes through a number of processes or stages before it reaches its customer. At each stage a transformation will be made which adds value to the product. Porter et al. (1985) used the term *'value chain'* to describe the sequence of activities that create a product's value, and 'value chain analysis' has become a standard way by which strategic planners search for competitive advantage in their operations. An *internal* value chain is the series of transformations carried out upon a product within a single organisation. An *external* value chain is where a product passes through a series of organisations before it reaches the customer (see Figure 2.10). Typically a manufacturing organisation will have upstream linkages with its suppliers and downstream linkages with organisations in its distribution channels. Commercial sector firms compete by the efficiency of the value chain of which they are a member. A company which is part of an efficient value chain has a competitive advantage over a rival which is locked into a less efficient chain.

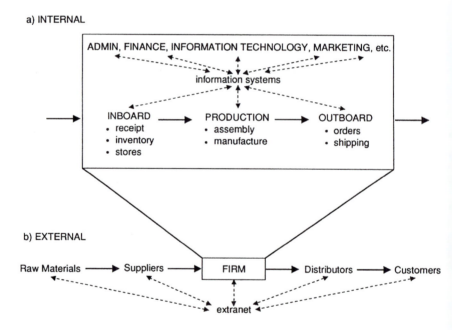

Figure 2.10 Internal and external value chains.

The value chain concept is useful because it helps managers visualise a production or service delivery process as a single entity, regardless of boundaries between sections or companies, and drives home the truth that the chain is only as strong as the weakest link. In the manufacturing sector, Japanese companies, in particular, have gained considerable efficiencies by insisting upon close relationships with their suppliers. Companies such as Toyota and Nissan are very willing to send teams of their own engineers and managers into supplier companies to sort out bottlenecks which they believe to be adversely affecting their value chains. Increasingly extranets are being used to bind external value chains more efficiently together, whilst intranets and other information systems co-ordinate internal value chains.

The relevance of value chain analysis to IT/IS planning is that investment in IT can often be used to improve the performance of a value chain. Imagine for example that a Chief Executive of a local authority is concerned that insufficient planning applications are being cleared, or that residents (customers) are having to wait too long for site enquiries to be returned. One way of attacking the problem would be to analyse the chain of activities that are necessary to produce the service and, if appropriate, invest in IT to resolve the bottleneck. GIS can be a device by which value chains that involve processing spatial data can be made more efficient.

CONCLUSION

This chapter has reviewed some of the key aspects of current thinking about the role of IS/IT in organisations. We have seen the growing insistence that corporate needs should lead investment in IT. We have also looked at the processes and tools that organisations might use to determine what IS/IT developments are required to underpin their missions.

GIS specialists should be aware of the concepts discussed here because corporate GIS proposals will need to fit into and be approved by the strategic IS/IT processes of host organisations. Simply pointing to the technological wonders of GIS won't do the trick, we need to be able to prove the business case for GIS. We referred earlier to the need for hybrid managers in business who understand both the technology and business domains within their organisations. The conclusion for this chapter must be that the GIS industry needs to produce *hybrid GIS managers* who understand organisations as well as the technicalities of GIS and, indeed, that all GIS practitioners should understand something of the business environments within which they work.

SELF-CHECK QUESTIONS

Management studies must rank second only to Computer Science for the amount of jargon it generates, and in this chapter we have introduced a number of new terms.

1.   For your own satisfaction see if you can define for yourself the following:

    — efficiency benefit
    — effectiveness benefit
    — competitive advantage
    — Transaction Processing Systems (TPS)
    — *Management Information System (MIS)*
    — Decision Support System (DSS)
    — Spatial Decision Support System (SDSS)
    — Enterprise Information Systems
    — Intranets/ Extranets
    — Virtual Organisations
    — SWOT
    — a 'cow', 'dog', 'star' and 'wild cat'
    — internal and external value chains

2.   Draw and annotate an organisational triangle:

    — what are the characteristics of the information required at each level?
    — what types of system are required at each level?
    — what roles might GIS play at each level?

WHAT YOU HAVE LEARNT IN THIS CHAPTER

  *   *the need to be aware of the strategic nature of IS/IT planning and investment*
  *   *the generic benefits offered by investment in Information Systems —*
      efficiency, effectiveness, competitive advantage
  *   *the generic roles of Information Systems within organisations*
          — organisational triangle, TPS, MIS, DSS, EIS, SDSS
          — how GIS fits within these generic roles
  *   *the emergence of intranets as technologies leading towards 'Enterprise' information systems*
  *   *the strategic planning process*
          — Business strategies, mission statements, objectives, CSFs, PIs
          — IS/IT strategies
   *   *tools for strategic thinking*
          — SWOT analyses, Boston Matrix, Value Chain Analysis

GOING FURTHER

There is a very large literature about the role of IS/IT as a strategic element within business planning. In preparing this chapter the following texts were used most frequently:

EDWARDS, C., WARD, J. and BYTHEWAY, A., 1991, *The Essence of Information Systems*, (London: Prentice Hall).

SILK, D.J., 1991, *Planning IT: Creating an Information Strategy*, (Oxford: Butterworth-Heinemann).

Both these are relatively short and crisp introductions to information systems management. Silk adopts the interesting approach of assuming the reader is a manager trying to create an information strategy and then explaining the sequence of activities which the reader should go through to be successful. Anyone who is trying to pilot a new GIS project through their organisations would benefit from reading Silk.

Of the many books which consider the impacts of computer technology upon organisational forms, the one directly referred to in this chapter is:

MARTIN, J., 1966, *Cybercorp: The New Business Revolution*, (London: McGraw-Hill).

Chapter Three:
GIS Systems Development Methodology

In this chapter we explore aspects of the 'typical' methodology by which organisations have attempted to determine whether GIS could be justified. We consider:

* *what is an Information Systems Development Methodology*
* *a 'typical' GIS ISDM*
* *a number of frequently used GIS development techniques*
* *a critique of the typical GIS ISDM*

'It is my experience that people (i.e. organisations) that are interested in GIS ask all the wrong questions — they want to know what's the best software, what's the best platform, etc. When I ask, "what is it you want to do?" the response is usually a lot of head scratching followed by "make maps" (or some other really vague remark). So, it is no surprise that determining user requirements can be a long (and sometimes painful) process — kind of reminds me of my kids who never want to do their homework either...'
Scott Freundschuh, e-mail, GIS-L, 1993.

INTRODUCTION

In the last chapter we looked at the strategic issues associated with the roles of Information Systems within organisations. In this chapter the focus is tighter as we look specifically at how organisations have attempted to justify introducing Geographical Information Systems and at the sequence of activities which have customarily been followed by organisations in order to implement a GIS.

Imagine you have been appointed leader of your organisation's GIS working party with a remit to investigate your need for GIS and, if required, overseeing its introduction into your organisation. What would you do? What would be your plan of action? Who would you want to talk to? What techniques would you use? In this and the next two chapters we review some answers to these questions.

In this chapter we draw heavily upon the specifically GIS literature, our intention being to build a generic GIS development methodology based upon published descriptions of GIS implementations and upon recommendations in GIS texts. As we will explain, however, the generic GIS development methodology ends up looking very similar to the traditional 'Waterfall' systems analysis model which was developed in mainstream computing in the 1970s and which has been widely criticised in the modern computer science literature. Accordingly, in Chapters Four and Five we will look at some of the more sophisticated Information System Development Methodologies (ISDMs) which have developed in mainstream computing. These alternative methodologies have not, as yet, been widely adopted by GIS projects but they do contain ideas with which we believe it is important GIS professionals should consider and which should help improve the success rate of future GIS projects. There appears at present to be a divergence between practice and theory with regard to GIS implementation strategies. Published accounts of actual GIS implementations seem, consciously or not, to have followed the traditional, technology-focused, 'systems analysis' methodology described in this chapter, whereas GIS academics have increasingly argued in favour of the more sophisticated approaches discussed in Chapters Four and Five. (Grimshaw, 1994; Marble, 1995; deMers, 1996).

WHAT IS AN INFORMATION SYSTEMS DEVELOPMENT METHODOLOGY?

As a preliminary, we need briefly to discuss what is meant by an Information Systems Development Methodology (ISDM).

Informally, an ISDM is a tool kit of ideas, approaches, techniques and tools which systems analysts use to help them translate organisational needs into appropriate Information Systems.

Definition

An **ISDM** is:
'... a recommended collection of philosophies, phases, procedures, rules, techniques, tools, documentation, management, and training for developers of Information Systems.'
Avison and Fitzgerald, 1988.

ISDMs vary considerably, but it is possible to highlight some general characteristics:

Philosophy: Avison and Fitzgerald (1988) make the essential point that methodologies need to be based upon a *philosophy* — a fundamental viewpoint about what an Information System is, and how such systems should best be created. This point will be taken up again in Chapter Four where it will be seen that development methodologies differ widely in their underlying philosophy. Some methodologies are clearly based upon an engineering metaphor — designing an Information System being regarded as being a similar kind of problem as designing a bridge or machine. Other methodologies draw more heavily upon social science paradigms and regard designing an Information System as primarily an organisational challenge.

Stages: Most methodologies consist of stages, each of which will itself consist of sub-stages. Each stage, and sub-stage, contains checklists and prompts to guide the developer towards a successful outcome. It seems that developing an Information System is such a complex problem that some form of 'top–down' approach, wherein an initial high level view of the problem is progressively exploded into a series of more detailed local views, or stages, is essential.

Techniques: Most methodologies have *techniques* embedded within them. Techniques such as Cost–Benefit Analysis, Prototyping and Benchmarking are commonly found in the early, feasibility stages of development methodologies. In subsequent data and process modelling stages, specialist analytical tools such as Normalisation, Entity-Attribute-Relationship, Data Flow Diagrams, and Action are often employed.

Tools: An Information System development *tool* is a piece of software that has been developed to make using a technique easier. For example, in the UK the Local Government Management Board's Geographical Information Advisory Group (LGMB, 1992) created a spreadsheet model specifically to assist local authorities to carry out GIS Cost–Benefit Analyses. For many data flow and process analysis techniques there are software packages which automate aspects of the techniques. Typically these packages might draw the techniques' diagrams, check the logical consistency of the model being created and, in some cases, automatically generate data dictionaries and/or computer code. There are also numerous project management tools designed to aid the phasing of implementation tasks

On a grander scale, a number of *'workbench'* software systems have been developed to support an entire methodology through each of its stages. The acronym *CASE* (Computer Aided Software Engineering) describes the use of integrated software to support an Information Systems Development Methodology, with the software packages themselves being known as *'CASE tools'*. The use of case tools can considerably shorten, and make more flexible, the Information Systems development process, and because of this such tools have begun to impact upon modern ISDM thinking. (We return to this point when considering alternative ISDMs in Chapter Five.) Although they are not yet as well developed as in the mainstream Information Systems industry, GIS CASE tools are emerging which will make more efficient the GIS design and implementation cycle (Kendrick and Batty, no date).

Graphics: Most ISDMs use graphical forms of problem representation, since graphs and diagrams are capable of conveying large quantities of information in a concise and simple form. Also diagrams can be used as a means of communication between the analyst and users.

Scope: ISDMs differ considerably in their degree of comprehensiveness. Some methodologies attempt to cover the entire development life-cycle from inception to maintenance, others are focused upon only particular phases or aspects of the cycle.

A 'GENERIC' GIS DEVELOPMENT METHODOLOGY

Time now to see what sort of ISDM philosophy emerges from the pages of the GIS literature.

Journals such as *Mapping Awareness* carry many descriptive accounts detailing how various local authorities and utilities investigated their need for GIS and how they subsequently implemented their GIS strategy. Similarly in the academic GIS texts there is usually a chapter about how to introduce GIS in organisations. Bodies such as the Local Government Training Board (1992) and the Royal Town Planning Institute (1992) have produced their own guides to how it should be done. Each of these accounts differs in detail, but they do seem to share a common approach. Usually these accounts take the form of a 'waterfall' model, with each stage leading seemingly inevitably and precisely to the next. Often there is a flow diagram to illustrate the model. (The RTPI's methodology, for example, starts with a box on the left labelled 'PUT GIS ON AGENDA' and ends on the right after twenty or so stages with a final box labelled 'USE SYSTEM'.) Most of the published GIS methodologies also share similar techniques — Cost–Benefit Analysis, User Requirement Studies, Data Flow Analysis, Pilot Studies and the like are repeatedly mentioned. Figure 3.1 outlines a generic GIS methodology which contains the stages and techniques which are often mentioned in the GIS implementation literature.

Anyone who has ever been involved with any form of planning will understand that diagrams such as Figure 3.1 need to be taken with a large pinch of salt. Planning, in reality, is a much more messy process than is implied by such diagrams. Things that should be done in sequence get done in parallel, things get done in the wrong order, and some things don't get done at all. Sometimes the entire process is driven backwards — you know what the outcome has to be and the planning process is followed simply to provide a justification for the predetermined conclusion. In many ways the methodology diagrams presented here and in the textbooks are like the physics experiments one used to write up at school — after the fact rationalisations of how it would have been done if it had been done properly. Diagrams such as Figure 3.1, however, do provide frameworks around which authors can explain their ideas about how Information Systems should be developed.

We will now work through the stages of the composite GIS methodology.

Another GIS 'Joke'

Question: 'Why do people buy a GIS?'
Answer: 'Because their neighbour has one.'

Newell, Smallworld Systems, quoted in Gunther, 1997.

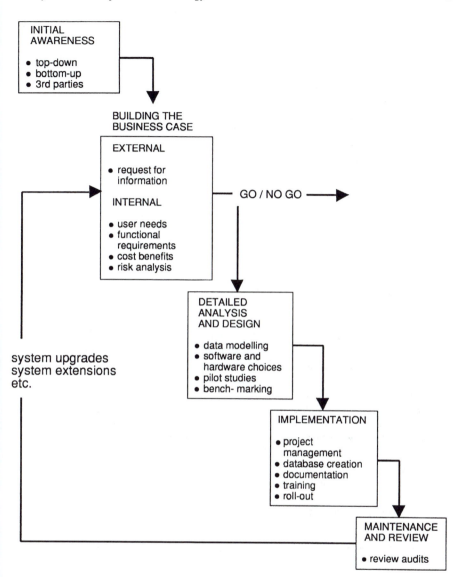

Figure 3.1 A generic GIS development methodology.

Initial awareness

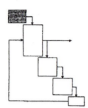

Initial interest in an Information System can stem from a number of sources. *'Bottom–up'* pressures for change develop when employees become frustrated with the way they presently do their jobs. With regard to GIS, Aronoff (1989) suggests a number of problem areas that might prompt officer dissatisfaction with existing manual methods within a local authority:

— Spatial information may be poorly maintained, so that maps and property lists are seriously outdated and questions cannot be reliably answered without rummaging through stacks of paper records (Gossette, Ferguson and Dueker, Unit 60, 1990).
— Spatial data are not stored in standard formats, so that the accuracy of mapped data varies across a city's area; several departments collect and manage the same spatial data.
— Data are not shared between departments because of concerns about confidentiality and or legal restrictions. Because possession of information conveys power, data transfer may be inhibited because of internal 'political' considerations.
— The 'CFTM' (Can't Find the Map) factor, may mean that delays caused by manual retrieval and processing of spatial data become intolerable.
— New requirements emerge within the organisation which traditional methods of handling spatial data cannot fulfil.

Officers working on a day-to-day basis with mapped information will often be keenly aware of the bottle-necks caused by manual mapping methods, and frequently middle managers, frustrated by the limitations which their groups are suffering, become 'lobbyists' for GIS. Where there is a ground swell of support for change from the middle of an organisation the prospects for subsequent success are increased, as there will be an inbuilt willingness to accept new technology. On the other hand middle managers are unlikely to be fully aware of the resource constraints and strategic issues which might argue against proceeding.

Pressure for change will often come *from the top* as strategic managers see the benefits that would flow from an integrated approach to spatial data handling. The advantages of top–down awareness are that senior managers will have the power to direct funds and organisational change to promote GIS. The disadvantages of top–down promotion are that senior officers are unlikely to be either fully aware of the complexities of operational procedures or fully conversant with technical aspects of GIS, so that their proposals might involve inflated expectations and unrealistically short deadlines. There may also be the problem of resistance from staff.

Third parties can also provide a stimulus towards new technology. Organisations are often influenced by the special interest groups and professional organisations to which their officers belong. With regard to GIS, bodies such as the Royal Town Planning Institute, the Local Government Management Board, and the

British Computer Society have GIS interest groups to keep members informed of developments. Organisations are sometimes effectively forced into adopting new technology by changes made by external organisations. In the UK, for example, the Public Utilities Street Works Act (PUSWA) required that local authorities and the utilities exchange large volumes of information about street works. Some utility companies are now well advanced in their use of GIS and will want to pass spatial information to local authorities in digital form. Any local authority which does not invest in some form of digital capability to receive messages from the utilities could soon find itself in difficulties. Similarly the UK's Care in the Community legislation which requires local authorities and health authorities to share information about the locations and needs of clients much more closely than has previously been the case, has created pressures on both sides to reassess their Information Systems. Cambridge Consultants (1993) also point to the role that specific events can have in raising awareness of the value of GIS, citing the raised awareness of GIS in the armed forces consequent upon the successful use of GIS in the Gulf conflict and the significant role that GIS played in assessing the Exxon Valdez oil spill. More prosaically, Campbell and Masser (1995) indicate the importance of the normal mechanisms of commerce in arousing interest in GIS, pointing out that in two thirds of their sample of UK local governments, vendors had been responsible for initially alerting staff to the benefits of GIS technology (and in most of these cases the local authorities concerned eventually bought from these same vendors).

Building the Business Case

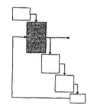

If the pressure to introduce GIS is strong enough to persuade senior managers that the proposal warrants further investigation, the next step usually will be to appoint a 'GIS Project Group' or 'Working Party' to carry out feasibility studies. In our generic methodology we've called this stage 'Building the Business Case'. We'll subdivide the wide range of activities into two broad activities:

- Scanning the external environment;
- Internal investigations.

Scanning the external environment

Much can be learnt from the success (and failures) of pioneer organisations and from contacts with potential hardware, software, and data suppliers. Contacts with external bodies can shorten development times and can help avoid repeating the costly mistakes of others. A danger to be avoided is to focus too quickly upon issues solely within the organisation.

Among the many topics upon which a GIS Working Party might seek external advice, the following could be important. What has been the experience of similar organisations which already have GIS projects — do they regard their GIS projects as 'successes'; were they completed within expected timescales and within budget?

Which were the most troublesome aspects of the development process? Do organisations which already have GIS projects similar to the one which is proposed have clear views about appropriate software and hardware platforms? What are their views of the support they have had from their vendors? Do the relevant professional bodies offer guidelines with regard to GIS? What is happening with regard to digital data from the national mapping agency and other major data suppliers? If the project will require a large data conversion exercise, what services might be available from specialist data conversion firms? What is central government's attitude — could legislation alter the case for GIS?

Many GIS project groups appear to go about the task of gathering information from the external environment in fairly informal ways — attending conferences, inviting demonstrations from vendors, reading relevant material, making visits to reference sites, etc. A more structured approach would be to send out a series of questionnaires to relevant organisations. In practice, this latter approach appears most often to be used with respect to obtaining information from vendors, by means of a *Request for Information (RFI)*. Essentially the RFI is a questionnaire sent to all relevant vendors which is designed to elicit information not only about the technical aspects of a vendor's products, but also about their ability to provide support and their presence in the market-place (Figure 3.2) (Gossette, Ferguson and Dueker, Unit 60, 1990).

Request For Information

An RFI might ask about:

- **General company information**
  — company history, corporate status, profitability, ancillary products

- **System capabilities**
  — functionality, modularity, customisation capabilities, development paths

- **Hardware and Software requirements**
  — operating systems, hardware platforms, networking capabilities

- **Customer references**
  — number of established sites, number of sites locally (e.g. within UK), number of sites in similar businesses, reference sites (i.e. available for inspection)

- **Customer support**
  — training programmes for new customers, maintenance and support systems, numbers of locally based support specialists

- **General pricing information**

Based on Gossette, Ferguson and Dueker, 1990.

Figure 3.2 RFIs — Requests for Information.

Rather than each individual local authority or utility having to prepare its own RFI, there is scope for umbrella groups to produce generic RFIs for their members. Thus the Geographic Information Steering Group (GISG) of the UK Local Government Management Board (LGMB, 1992), for example, have published a 55 page specimen questionnaire which they suggest can provide a template for local authorities wishing to obtain details of GIS products. (A danger with specimen questionnaires is that they can be unthinkingly used. Vendors in the UK are now complaining that they are often asked to fill in long and familiar questionnaires for the simplest of applications and where it is apparent that many of the questions are irrelevant. In extreme cases, the labour cost to vendors of filling in the questionnaire can exceed the price of the software!)

It will often be the case that many of the members of a GIS project team will not have had direct experience of GIS. They may be enthusiasts for GIS but probably not experienced users. In these circumstances the preliminary researching of the external environment, and particularly the RFI, will not only provide specific information for internal reports but also form an essential early part of the learning curve for the team.

Internal investigation

Although scanning the external environment will provide a context for a project team's work, the major emphasis will be *internal* — trying to establish the internal case for GIS.

Under this heading we will look at three activities:

a) User Needs Analysis
b) Cost–Benefit Analysis
c) Risk Analysis.

The first two of these appear very frequently in published accounts of GIS developments. Risk Analysis gets less attention than we think it should.

a) USER NEEDS ANALYSIS

> 'The User Requirements Analysis stage is the most important in the entire lifecycle and involves the creation of a specification to correspond with the organisation's business plan, skill base and resources.'
> McLaren, 1992.

Almost all the published accounts of GIS development strategies refer at some point to a **User Needs Study (UNS)**. Essentially what this means is that the project team goes out into the organisation and by means of interviews, questionnaires, observing existing work processes and analysing data flows, attempts to determine the demand for GIS.

The depth and formality with which project teams conduct User Needs Studies varies considerably from organisation to organisation. Some organisations have adopted very informal methods to gain a 'first cut' impression of support for GIS. Sometimes, however, more formal methods may be used with the intention, not merely of proving the demand for GIS but also, looking ahead, to provide a secure starting point for formal data design. The New York State *GIS Development Guide* (Becker et al., 1997), for example, pursues this latter approach, providing template pro-formas by which data-flows, entity descriptions and application descriptions can be recorded in detail. In the UK, the Local Government Management Board's (LGMB, 1995) 'GIS: Go with the Flow' project which attempts to develop a generic model of data-flows and process-requirements within areas of local government activity was similarly based upon formal analysis of requirements.

Although UNSs appear typically to be conducted primarily by means of interviews there are a couple of graphic techniques which are often used to help clarify issues within UNS reports:

Grid Charts: These can be used to highlight both aspects of the current position with regard to use of spatially referenced data and aspects of the envisaged situation should a GIS be introduced. Figure 3.3, for example, shows two grid charts which might result from a local authority UNS. Figure 3.3a shows the current interaction of spatial data sets with user departments and also indicates the intensity of use. Figure 3.3b has similar axes but suggests what the position would be once a GIS is in place, indicating which departments would have responsibility for contributing particular data sets to the corporate system. Grid charts, although very simple, can be very revealing. Looking across the rows reveals which data sets are used by many departments, and which therefore might be early candidates for inclusion in a corporate GIS. Looking down the columns indicates which departments are heavily dependent upon spatial data sources, and which departments are not major users of spatial data. In some grids it may be possible to identify clusters of departments which use similar data sets and which might provide a focus of a shared GIS application. In local government, for example, it is likely that those departments involved in the land development process will share common data sets.

Data Flow Diagrams: Informal data flow diagrams can be used during a UNS to provide a first view of the flows of spatial data within and between departments in an organisation (Figure 3.4).

Data Flow Diagrams can help members of the UNS team get a mental grip on the often complex structure of information and map flows within an organisation. They can also provide a consistency check for the information derived from interviews. If the Economic Development Unit has told you during its interview that it receives its basemaps from Planning, then when the Planning interview is undertaken there should be mention of a basemap flow from Planning to EDU. A line on a data flow diagram that has a destination but no origin can be something of an embarrassment.

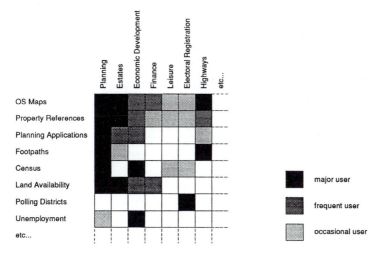

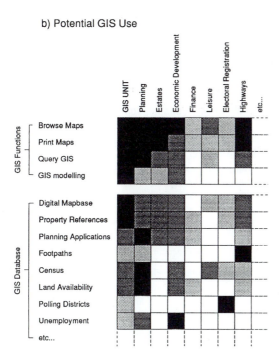

Figure 3.3 User need grid charts — for a local authority. (Source: modified from Bromley and Coulson, 1989, *Geographical Information Systems and the Work of a Local Authority*, Dept of Geography, Swansea University. Used with permission.)

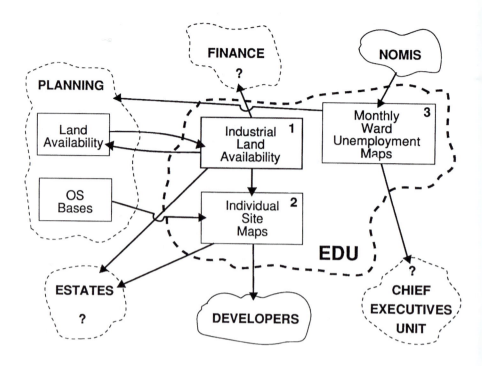

INTERVIEW NOTES:

Processes **1** Colours industrial sites red on bases supplied by planning.
Twice a year.

2 Obtains larger scale OS bases from planning technicians.
Often scale / edge problems.

3 Ward unemployment statistics received monthly by modern link
to NOMIS (National Online Manpower Information System).
Mapped using Atlas. N.B. Poor quality reproduction.

? Interviewee not sure of source destination details. CHECK!

Figure 3.4 Data flow diagram, e.g. From an economic development unit interview.

The end product of a UNS should be a report that gives an organisation, at
least, a preliminary basis for deciding whether to proceed further with GIS. With
varying degrees of formality, the organisation should gain knowledge about four
broad aspects of its GIS requirements (Ferguson, Unit 61, 1990):

Spatial Product Requirements: What spatially based information products
does the organisation currently produce, how often they are produced, who

produces them, how are they produced, how satisfied are users with the products, what bottlenecks are there? Also, the UNS can look into the future — what spatially based information products would decision makers *like to have* available which are presently not produced, e.g. new kinds of analysis, analyses at a finer spatial scale or more frequently produced, using more current data, etc.?

Spatial Data Sets: An inventory of the spatial data held by the organisation — what spatially referenced data sets, including maps, does the organisation hold, who is responsible for obtaining, maintaining, and disseminating the data, what level of duplication of data is present, how are the data held (tape, disk paper, etc.), what are the volumes of data, what security and back-up procedures are there, which departments are the key data holders within the organisation? What spatial referencing systems are used within the organisation (UPRNs, post-codes, centroids etc.). Are there common data standards? What spatially referenced data sets does the organisation currently *not* have access to — what are the gaps in the organisation's present data sets? How does the organisation interact with relevant external organisations with regard to transfer of spatial data?

Preliminary Specification of GIS Functionality: Which GIS functions would be needed to produce present and required information products? Which functions are common to all products (e.g. plotting)? Which functions would be specialist requirements limited to only one or a few products?

Attitudes towards GIS: Which departments/officers would support expenditure on GIS? Which are aware of possibilities GIS would offer? Which group might provide an appropriate site for a 'showcase' application, etc.?

Although UNSs are very widely used in GIS development methodologies, there are a number of practical and conceptual difficulties associated with conducting a UNS. These include:

The composition of the UNS team: Who should conduct the interviews? Ideally the interviewers need to know both about GIS and about the functional areas of responsibility of the interviewees. As persons with such dual knowledge are likely to be rare, UNS are sometimes conducted by pairs of interviewers, one being the technical specialist (perhaps an outside consultant), the other the functional area specialist. Both need to have interviewing skills. Double teaming, of course, greatly adds to the costs of UNSs.

Areas of the organisation to be: Within a local authority, some areas would be obvious targets for interviewing — Planning, Economic Development, Highways, Estates etc., but should you also interview less likely areas? What about Chief Executives, or Finance, or Legal or Leisure Services? If you do interview these areas you run the risk of wasting a considerable amount of time (and money) on fruitless interviews. If you do not interview them, you run the risk of missing out on some worthwhile GIS applications which might be lurking in apparently unpromising areas.

Who should be interviewed: Technical staff, that is the people who prepare information products and who live with the day-to-day realities of data processing, are likely to be able to provide operational details of spatial data processing and are often likely to be aware of the potentials of technology to improve their performance. Technical staff, however, are unlikely to be able to judge the importance of the information products they produce and are unlikely to be aware of the organisational and budgetary constraints which might temper enthusiasms for investment in GIS. So perhaps interviews should be targeted upon the decision makers within the organisation, as these people will be able to take an informed view of just how valuable having improved spatial information products would be to the organisation. But such higher level personnel may be unaware of the technology and also unaware of the realities of the data processing which goes on below them in the murky depths of their organisations.

Limited awareness/Unrealistic expectations: A difficulty with UNSs is that interviewees may well only have a very limited awareness of GIS and so their answers may underestimate the potential contribution of GIS to their activities. If a person does not know of the availability of OS digital maps, she cannot be aware of their potential contribution to her tasks. Sometimes being too close to traditional ways of doing things limits the vision of what is possible with new technology. Existing expertise and expectations become blinkers.

Limited Awareness

'The original spec for plotting was drawn up by Drawing Office staff who are used to producing high quality work. They specified plotters that produce high grade, non-degradable maps at the cost of £1 per A3 sheet. Actually, now GIS technology is available, the operational staff prefers 'throw away' copies and the company is buying Inkjets.'
Source: Diploma student.

Conversely, officers may well have read some of the more fanciful literature about GIS and thus have unrealistic expectations about the capabilities, costs, and timescales involved in GIS. In practice UNS exercises often have an educational aspect in that to be worthwhile interviewees need to be provided with a balanced understanding of the potentials and limitations of GIS before being asked questions about the utility of GIS for them. Often introductory seminars and papers are produced to raise awareness prior to UNS interviews.

Excessive change/Too little change: Chrisman (1987) provides an interesting early critique of the UNS process. He observes that UNSs having typically adopted a social science questionnaire/interview approach tend to produce results which allow considerable scope for interpretation by the project teams involved, and that there is a danger that the evidence will be used either to recommend too much or too little change. Duplications of data holdings that from an analyst's viewpoint might seem ripe for rationalisation may often have deep-seated human and organisational justifications. On the other hand if the UNS team is too sensitive to the nuances of existing practices, the analysts can end up recommending a GIS which simply replaces a chaotic manual system with a chaotic computerised one. (The debate about the degree to which the introduction of new Information Technology should be used as an opportunity to sweep away previous work practices has risen to prominence in recent years with the growing interest in **Business Process Reengineering** and we return to this issue in Chapter Six).

Clearly conducting a User Needs Study is a very imprecise and problematic task, but it is one which seems to be essential. A UNS will often be beneficial to organisations, regardless of the final outcome, in that such studies force organisations to reconsider the adequacy of map processing procedures which may have grown in an ad hoc manner over many decades. Even if the organisation decides not to proceed with a GIS development, a UNS may still prove to have been a revealing exercise, and lead to improvements in manual spatial data systems.

b) COST–BENEFIT ANALYSIS

At any time managers responsible for developing a company's IS strategy will almost certainly have more proposals in front of them than they have the resources to support. At the same time as a local authority's GIS project group is developing its plans, there might, for example, be a need for a new personnel Information System, the highway engineers might be threatening deaths on the road if their new system isn't progressed, the education department could be demanding a new system to cope with the latest bright ideas from central government and so on. The Information Strategy group, therefore, needs some means of determining which of a number of competing possibilities will provide the greatest net benefits for the organisation. A technique which is often used in these circumstances is **Cost–Benefit Analysis (CBA)** as this technique purports to render costs and benefits onto a common basis and thus to make possible direct comparisons of the net benefits which would flow from widely disparate proposals.

Virtually every GIS project seems to have been subject to some form of Cost–Benefit Analysis. Indeed, of Cost–Benefit Analyses of GIS proposals are so commonplace in the local authority sector that, as noted above, the UK Local Government Management Board published a spreadsheet CBA model specifically to help local authorities to evaluate the value they would derive from investing in GIS (LGMB, 1992). A curious thing about CBA, however, is that although almost every project team feels obliged to use it in some form or another, everyone who has actually used it will be acutely aware of its limitations.

The basic idea behind CBA is straightforward. All the costs which a system would generate over its lifetime are accounted for on a consistent monetary basis. Similarly all the benefits which would accrue from the system over its lifetime are allocated a money value, and thus it should be possible to calculate a net benefit–

cost ratio. The insistence upon costs and benefits being expressed in monetary units is important as, in theory, it gives decision makers a consistent basis upon which to compare the relative merits of otherwise disparate proposals — a GIS proposal as opposed to, say, a highways or educational IS system. In a simple world, the IS decision makers should simply choose the proposal which offered the greatest benefit/cost ratio. In practice, however, producing believable and comprehensive entries to a CBA balance sheet has proved very difficult indeed.

COSTS

Of the two sides of the CBA balance sheet, identifying costs should be the least problematic, primarily because costs are likely to have direct monetary values. In practice, however, evaluating the costs associated with a GIS project is not always straightforward. The number of cost components which need to be included may not be immediately apparent, and for those which are identified it may sometimes be difficult to provide an exact monetary value ahead of time.

Typically the costs associated with a GIS proposal might fall under the following headings:

Procurement Costs: In a full accounting system the costs of investigating the need for a GIS should be included. The work of the project group, costs of any consultancies, travel, pilot projects, and benchmarking costs, etc., will involve an organisation in expenditure which should be counted against the project in the final balance sheet.

Start-up Costs: These are the 'headline' costs which are most often noticed, as they are the big, capital outlays which have to be sanctioned by managers at the start of a project. Start-up costs are likely to include the costs of hardware and software purchases; costs of modification to buildings (air conditioning, security, cable ducting, etc.); network costs; costs of searching for and hiring project staff; staff training, etc. Where loans are taken out to cover start-up costs, the interest charges on these loans must also be included.

Data Conversion Costs: Although attention tends to focus upon initial hardware and software costs, experience of large GIS projects so far suggests that such costs will actually represent less than 20% of total cost over a project's life-cycle. Indeed, for many large GIS projects, the costs associated with data conversion alone are likely to dwarf hardware and software costs. Creating a worthwhile and accurate GIS database for an operational system, for example, converting a local authority's manual Terrier system to GIS, or establishing a GIS database for a utility, is a mammoth task which will soak up many months of labour costs. Conventionally it is estimated that database creation will account for around 80% of the total cost of a GIS project. In one exceptional case, Campbell and Masser (1995) report a local authority GIS system in which data costs were believed to have been sixteen times greater than expenditure on software and hardware. The size and cost of the data conversion task appears often to be underestimated in GIS projects.

Other Data Costs: In addition to the cost associated with converting the internal database, there is likely to be a significant cost associated with purchasing external data sets — OS digital maps, property geo-referencing systems, postcode systems, etc.

Maintenance Costs: Start-up costs are by definition once-off costs, but clearly there will be recurring costs associated with any project. Software and hardware maintenance and upgrades will represent a major recurring cost. With vendors typically charging around 10% of initial purchase price as an annual maintenance charge, and pushing regular upgrading, the recurring costs of hardware and software can exceed the initial purchase prices after not too many years. Korte (1996), for example, suggests allowing for hardware to be replaced every four to six years. There will also be recurring data costs. Also any project will incur not only the salary costs of staff employed on the project, but also the continuing overhead costs of supporting them — consumables, office equipment, telephones, floorspace etc.

Alter (1992) suggests that a common failing in CBA of Information System proposals is to fail to cast the net wide enough when considering costs. He observes that many items such as the amounts of management time devoted to setting up a project and to subsequent troubleshooting tend not to be accounted for. With regard to post installation tuning and trouble shooting, Lay (1985) similarly makes the point that these items may impose significant costs but that they are difficult, if not impossible, to quantify at the pre-planning stage. The potentially significant costs of 'bedding-in' a system are often, therefore, omitted from CBA estimates.

Conversely, it is possible to exaggerate costs by not anticipating cross benefits from costs. If, for example, a workstation is bought as part of the set-up of a GIS project but is also subsequently used to run the department's personnel system, how much of the cost of purchase should be entered against the GIS project? Many of the staff who run a GIS system will also be involved in many other tasks, so how can the staff costs associated with the GIS project be reasonably assessed? Much is often made of the high costs of database creation in GIS. But if the data would have had to be captured and recorded anyway, perhaps only any *additional* cost associated specifically with using GIS should be counted against the GIS project?

At first sight identifying costs might seem straightforward, but in practice things are more difficult. Arriving at a balanced cost estimate for a GIS project is not easy.

BENEFITS

If there are difficulties associated with identifying costs, the ambiguities associated with identifying and allocating a cash value to benefits are much greater. At least costs tend to have a natural cash value associated with them. However, how does one estimate the cash value of, say, the increased job satisfaction which employees might derive from the introduction of a GIS? Again, one approach is to try to break down benefits into a number of types. Here we will follow Antenucci et al.'s (1991) classification of potential GIS benefits:

Type 1: Quantifiable efficiencies in present practices: For many tasks for which a GIS will be used it should be possible to contrast the present costs associated with performing tasks manually with the anticipated cost of performing the tasks once the GIS has been established.

The net benefit here is obviously the difference between the two costs, i.e.

$$\text{Net Benefit} = \text{Present costs} - \text{GIS costs}$$

It should be possible, for example, to measure how long it takes for a map to be drawn manually, or an enquiry to be answered manually, and to calculate the staff costs associated with these tasks and then to compare these with the costs of equivalent GIS operations.

A related approach is to estimate *cost avoidances*, i.e. to identify those costs which a present manual system incurs which would no longer be necessary if a GIS were in place. Earlier we referred to floorspace savings associated with using Information Systems, and for GIS this can be a major item. Think of all the map tanks and map chests which a local authority will have across its offices. If a GIS is introduced, this expensive floorspace will no longer be required. Another significant cost-avoidance which derives from moving from paper to digital mapping is that the costs associated with manual redrafting no longer have to be borne. Many organisations maintain large paper map sets onto which operational data are overlain (e.g. planning applications, land holdings, highway schemes, etc.). Because such map sets are intensively used, they eventually wear out and need to be replaced — say once every fifteen years. Redrawing and checking land boundaries onto a single Terrier sheet can take a technician several days. Redrawing an entire set of Terrier sheets for a local authority is a major task, and a major cost, which is simply not required with GIS.

Calculations made by organisations to estimate the efficiencies which introducing GIS can provide, do make clear that for large users of topographic maps significant saving are indeed available from adopting GIS. England (1966), for example, reports that in UK local authorities work done with conventional mapping systems breaks down into the following categories:

Activity	Time spent
Look for, retrieve, return maps	45%
Extract information from them	33%
Copy all or part of maps	13%
Modify them	9%
	100%

and suggests that the introduction of digital mapping can at least halve the time spent on these activities. For his own authority of Gloucestershire, this would translate into an annual saving of £300,000 and result in a pay-back time of just two years. Smith and Tomlinson (1992) similarly suggest a 50% reduction in the time required to search and respond to requests for mapped information when access is provided via a GIS. Huxhold and Levinsohn (1995) cite productivity increases of between 25% to 75% gained on mapping and drafting tasks upon the introduction of GIS. Korte (1996) suggests that although 50% might be a reasonable estimate of the overall savings which an organisation might expect to make by transferring to digital methods, for some tasks timings can, in fact, be reduced still further and suggests a time reduction of 4:1 for many drafting and map publication tasks.

Type 2: Quantifiable expanded capabilities: In every organisation there will be tasks which would provide value or better service but which are not carried out because the labour input required is too great. User Needs Studies will usually reveal a list of such tasks, which employees recognise as being worthwhile, but which never get done because of pressure of other work. In one UNS for an English local authority, for example, several officers independently mentioned that they would like to plan service provision at Enumeration District level, but could not do so because of time pressures, and were therefore obliged to use Census Wards as for planning purposes. Similarly, the Library Service research officer knew that the routes and stopping points of mobile libraries should be reconsidered more frequently to reflect changing demand, but rerouting by hand was a very time consuming task. The introduction of GIS can allow staff to carry out such tasks which previously were too demanding of labour to countenance.

In terms of the generic benefits outlined in Chapter Two, the Type 1 benefits discussed here are *efficiency* benefits, where the GIS is used to make existing tasks more efficient. Type 2 benefits are *effectiveness* benefits as the GIS allows existing staff to perform additional tasks. Type 2 tasks are quantifiable because it should be possible to measure the labour that would be required to do each additional task manually.

Type 3: Quantifiable benefits from sale of new products or services: A GIS might allow new products to be developed which can be sold, or might improve the quality of an existing product or service so that a higher price can be charged for it. For example, one rationale which has been proposed for local authority land and property GIS is that, as such systems dramatically reduce the time prospective developers have to wait for site searches to be completed, it is reasonable to make increased charges for the improved service.

Type 4: Intangible, unquantifiable, benefits: Here we meet a fundamental difficulty with CBA. With effort and some ingenuity it may be possible to assign reasonable monetary values to benefits of the kinds so far listed. Many of the most significant benefits which derive from introducing a GIS, however, will typically be of an intangible and effectively unquantifiable nature. 'How do you measure saving the kid who did not get hit by the car, or the value of storm drains not being clogged because we know where they are?' (Forest, quoted in Antenucci et al., 1991). Just how does one place a monetary value upon the improved self-esteem, increased staff morale, greater job satisfaction, improved staff retention, etc., which might result from introducing a new system? Many such benefits will simply be serendipitous, emerging as the system develops, and as such cannot possibly be built into an initial cost–benefit analysis.

A particularly thorny issue here is to estimate the value to be placed upon improved information to an organisation. If a GIS is being used to provide information to help managers to take better decisions, how is it possible to place a monetary value upon the 'better information' provided by the GIS? This is a key and almost unanswerable question: *'What is a particular 'piece of information' worth to an organisation?'* Without being able to produce a believable quantified answer to this question, it is virtually impossible to 'prove' the benefits that many GIS investments will bring.

Academics have suggested a number of theoretical methods by which a cash value might be placed on 'information' (Litecky, 1981; Parker, 1988; Dickinson and Calkins, 1990), but in practice project teams often abandon the logic of CBA and simply provide text descriptions of expected unquantifiable benefits.

It should be noted that some of the 'benefits' described above might imply an increase in costs, rather than a cost reduction. Using a GIS to allow employees to provide a better service, or to perform more penetrating analyses could mean greater activity and hence possibly greater costs. In some organisations, such 'soft' benefits would be disregarded and only 'hard' benefits which result in a reduction in 'bottom line' costs would be considered. GIS project teams will need to understand the financial rules by which their particular CBA will be judged.

TIME and CBA

A feature of any significant project is that costs and benefits will occur over a period of time. Typically there will be high start-up costs but benefits may take some considerable time to phase in. This adds a further complication to CBA analysis, because organisations have not only to identify costs and benefits but also to decide how they are going to view them through time. How long is an organisation prepared to wait before pay-back begins? How should an organisation balance a project which promises a moderate return in the short-term against a project which promises large benefits but over a much longer term? What form of discounting should be adopted, etc.?

Projecting cost and benefit streams on a consistent basis represents a technical task, involving concepts such as 'Net Present Value' and 'Internal Rate of Return', the details of which we can safely leave to accountants and economists, but Figure 3.5 provides a schematic of a typical GIS cost/benefit time-path, comparing the cost of continuing to provide a manual service with the costs of a similar level of service provided digitally. The manual line is shown as rising gradually to reflect rising labour costs. The GIS cost line reflects high initial purchase and set-up costs but the line then subsides to represent the costs of running the established system. It may take a period of years before the costs of running the GIS system are less than those of the manual system and pay-back begins.

In summary, CBA often seems to be an attempt to quantify the unquantifiable and one suspects its role is often to provide a quasi-scientific rationale for decisions which are in reality subjective. Decision makers seem to feel more comfortable taking decisions upon the basis of pages of figures rather than upon an openly intuitive recommendation that 'we believe this to be a good idea'. This being the case, wise project teams will take care to provide pages of figures, even if they have limited belief in them: 'It may be politic to present a decision as if it had been adopted primarily as the result of conventional CBA, although it is accepted in private that this is not the primary justification' (Nahapiat, 1984). When GIS is being used primarily to automate existing manual tasks it may be possible to arrive at a reasonable estimate of costs and benefits in measured terms. When GIS penetrates the middle and upper layers of the organisational triangle, however, the rationale for CBA becomes very questionable.

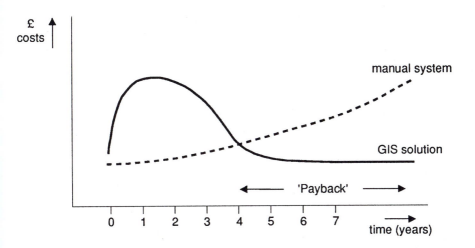

Figure 3.5 Projected manual system and GIS costs.

Interestingly, although in GIS developments the use of CBA is virtually ubiquitous, Silk (1991) reports that in a wider business context IS strategy makers are beginning to feel less need to cling to spurious numerical analyses. Silk asked 281 senior managers how their companies took IS investment decisions. He found that companies varied considerably in the extent to which 'hard' facts were required as a basis for approval. Some companies made some investment decisions as *'acts of faith'*, i.e. were prepared to back the belief of staff that a system was right for the organisation; some accepted a *'causal logic'* form of justification, where the beneficial consequences of investing in a system are explained without there being pressure for such benefits to be quantified. Perhaps as GIS becomes more established and decision makers more confident in its benefits IS strategy makers will be more ready to relax their reliance on quantification of benefits.

c) RISK ANALYSIS

A feature of CBA is that usually costs and benefits are calculated upon a broad assumption that the system will develop generally as planned, but as we saw in Chapter Two the history of Information Systems forces us to the rather gloomy conclusion that they often *do not* turn out as planned. They fail! Rationally, therefore, a project team should not only estimate costs and benefits on an assumption of things going broadly to plan, they should also try to estimate the risks of things going wrong.

Estimating risk is an interesting task. Alter (1991) suggests that one, informal, method could be to compare a proposed system against a hypothetical minimum risk system, which would have the following characteristics:

> 'A system to be produced by a single implementor for a single user, who anticipates using the system for a very definite purpose which can be specified in advance with great precision. Including the person who will maintain it, all other parties affected by the system understand and accept in advance its impact on them. All parties have prior experience with this type of system, the system receives adequate support, and its technical design is feasible and cost effective.'

The further a proposed system deviates from the minimum risk model the greater the risks. In most circumstances, GIS might be expected to contravene Alter's low risk criteria.

A more structured approach of risk assessment is to try to identify sources of risk and then score a proposal along a scale for each factor. McFarlan (1981) for example suggests that risks associated with Information Systems could usefully be divided into:

- *Size risk*, i.e. the more man-hours, the more staff, the longer the development time, the more departments involved, the greater the risk;
- *Structure risk*, i.e. the more organisational change required, the more managerial support demanded, the more job definition changes, the greater the risk;
- *Technology risk,* i.e. the more novel the hardware and software, the less previous experience of the development team, the greater the risk. Again, viewed by these criteria, you might conclude that most GIS developments are very high risk projects indeed.

Simply analysing the risks associated with a GIS project is clearly not sufficient, the objective will also be to *manage* and hopefully reduce them. Schofield (1996), for example, provides an account of how risks were managed during a major GIS data conversion project. Staff involved were required to enter risks as they were recognised into a 'risk register', using different coloured forms to indicate the severity of the risk. Regular risk management meetings were held to review identified risks and to plan appropriate remedial actions.

Detailed Analysis and Design

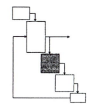

At the end of the 'Building the Business Case' phase of the methodology the project team presents its findings to the IS/IT committee and a decision is taken about whether to proceed further. If the decision is that the case for GIS has not been made sufficiently strongly to warrant further development, the methodology stops here.

If a decision is made to proceed, however, the nature of the methodology alters markedly. The purpose of the previous stages has been essentially to remove doubt about whether a GIS system is needed. Now a decision has been made to proceed, emphasis switches to developing the most appropriate GIS. In Boehm's (1981) terms the emphasis switches from *Validation* ('Are we building the *right* system?') to *Verification* ('Are we building the system *right?*').

Under the heading of Detailed Analysis and Design we will consider the following activities:

a) Data and process modelling
b) Benchmarking
c) Pilot studies

a) DATA AND PROCESS MODELLING

From the User Needs Study, the project team will have established an understanding of the products which the GIS will be expected to generate, the data which will be required, and the processes which will need to be carried out on the data. If the UNS has been conducted using a formal method, there may well also already exist some top-level data and process analysis material which can form a starting point for detailed analysis.

Once final approval has been given, however, it becomes necessary and worthwhile to focus upon the *detailed* design of the system. A detailed spatial modelling process will need to be carried out to determine which spatial entities are needed within the system and how each entity is to be represented and how each is to relate to other entities. For attribute data a similar sequence of conceptual and logical modelling followed by detailed specification will need to be followed. Analysis and decomposition of processes will be required.

Much of the work done within this stage of the methodology involves the use of detailed, sometimes software specific, techniques which it is not appropriate to pursue here. The output from this stage, however, should be a detailed specification of the required GIS system.

b) BENCHMARKING

Identifying appropriate GIS software represents a particular problem for many organisations. In some organisations GIS continues to be new technology and so there is not a pool of internal expertise to guide selection. Also, there is a wide range of software products, based on fundamentally different models, with widely varying functionalities, all masquerading under the name 'GIS', so that it is very easy to make a wrong choice.

Not surprisingly many organisations have chosen to tread very warily through the minefield of evaluating alternative GIS packages. (Although at the other extreme, we do hear whispers of organisations which have bought a GIS quickly because they happened to have money available at the time, and have worked out what they want to do with it later!)

Using information gained from Requests for Information sent to vendors, visits to reference sites and attending demonstrations, an organisation will probably by this stage have established a short list of products which are believed to be suitable. But how to choose from within the short list? In the published accounts two techniques — benchmarking and pilot studies — appear frequently as a means of making final software choices.

Anyone who reads the computer magazines will be familiar with an aspect of **benchmarking**. In these magazines there are often 'shoot-out' reviews where a number of similar packages are each given a series of 'benchmark' tasks to perform and their relative execution speeds are reported. The term *Quantitative benchmarking* is used where the results of the benchmark can be expressed as a numerical measure of performance. It is easy to see how simple quantitative benchmark tests for GIS systems could be devised. How long does it take different systems to overlay two standard polygon nets; how long does it take the same operator to digitise the same set of polygons using different digitising software, etc.? In practice, however, quantitative GIS benchmarking can become a complex task (Goodchild and Rizzo, 1987).

In *Qualitative benchmarking* the results of the evaluation are less precisely expressed. Sometimes, as in the computer magazines, a simple text description may be provided recounting the experience of the evaluators as they worked their way through standard procedures on different systems. Alternatively, a ranking system might be devised to position each GIS in order along scales such as user-friendliness, speed, and functionality, etc. The UK Local Government Management Group (LGMB, 1993), for example, provide a set of guidelines advising how local authorities should conduct GIS benchmarking exercises in which they recommend a detailed ranking and weighting system to aid selection. The box below outlines a much simplified ranking procedure, based on material in the Benchmarking Unit in the NCGIA's Core Curriculum (Goodchild and Kemp, Unit 63, 1990).

Qualitative Benchmarking

A ranking procedure:

a) Make a list of the functions which the User Needs Survey indicates will be required from the GIS system and allocate to each function a weight to reflect its importance to the organisation, e.g. 1 = desirable to 5 = essential.
b) For each system allocate the performance scores indicated below for each function:

    0   ABSENT
    1   ABSENT, but plans for implementation
    2   PRESENT, but slow
    3   PRESENT AND FAST, but difficult to use
    4   PRESENT AND FAST AND EASY TO USE

c) Multiply the desirability weights by the performance scores.
d) The system with the highest score wins!

In practice a benchmarking programme usually involves an organisation writing a *benchmarking script* which provides vendors with a detailed description of the required benchmark tasks. The organisation will usually also provide the test data sets upon which the benchmarks are to be performed. In this way benchmarking puts the organisation in control of the situation. The organisation specifies the tests and selects the data. This reverses the typical 'demo' situation where the vendor is in control and, understandably, is likely to emphasise the strengths of its software.

The implementation of the benchmarks will vary from instance to instance. Sometimes vendors may be required to bring their systems to the organisation's premises, sometimes company evaluators may visit vendor premises, and sometimes there may be a 'play-off' where two or more systems are simultaneously tested.

A GIS Grand Final

'In a GIS contest in Sweden in October 1992, Intergraph outstripped its competitors ESRI and IBM. The contest was part of the final presentation of results from a three-year government programme to enhance IT.

One of the problems the companies were asked to solve — in front of a live audience — concerned the real-time simulation of the effects of a gas leak on urbanised areas. Intergraph was the only GIS vendor capable of modelling the spread of toxic chlorine concentrations. The winning solution used as many as seventeen important variables, including wind speed and direction, gas pressure, turbulence factors and cloud density.'
Source: an Intergraph publicity handout — doubtless IBM and ESRI could point to their own successes.

Conducting a benchmarking programme can impose considerable costs both on the organisation and the vendors. The organisation has to specify appropriate scripts and produce realistic data sets. The vendors may have to commit staff for a considerable period to set up the tests. Often the vendors will expect the commissioning organisation to bear some of their costs. Clearly benchmarking, on a significant scale, can only be justified by individual organisations for major GIS proposals.

c) PILOT STUDIES

Pilot studies are frequently used as a means of testing the capacity of a particular GIS within the host organisation. Typically an organisation will arrange to hire the software for the period of the pilot study, with a provision that if a final decision to purchase is made the cost of the hire will be deducted from the eventual purchase price (a 'lease with option to purchase' contract). During the pilot test period, the organisation can thoroughly test the suitability of the software, and its assumptions about likely progress, by working up one or more of the key applications indicated by the User Needs Study.

A properly designed pilot study can provide considerable benefits for the GIS project team in addition to the primary one of testing the functionality of the chosen software:

- A pilot project can provide a means of demonstrating the potential of GIS to colleagues;
- A judiciously chosen pilot application might help gain the support of an influential person or department;
- It provides the project team with direct hands-on experience of GIS, possibly for the first time;
- It allows the project team to gain a feel for the type of company and type of support to which they would be committing;
- It allows assumptions about timescales and resources to be tested. Logically the best way to test assumptions about how long tasks will take and what resources will be required is, of course, actually to *do* the tasks.

A good pilot study considerably reduces the risks associated with introducing GIS into an organisation. If the pilot is 'successful' and there is a decision to go ahead with the pilot software, the in-house expertise gained during the pilot stage will speed the introduction of the main project. If the pilot 'fails' in the sense that the pilot software is deemed unsuitable, then better to discover this during the pilot than later with the main project. ('Successful' and 'fails' are deliberately in inverted commas here to suggest that a 'failure' can be a form of success if it causes an organisation to avoid making a wrong purchase. In drawing up the terms of reference for a pilot study management should be careful to stress that a decision not to proceed will be as welcome as a decision to proceed, and the officers involved in the pilot should not feel pressured to squeeze a 'success' out of a pilot at all costs.)

On the other hand, it is possible to point to a number of significant limitations of pilot studies as means of assessing the suitability of GIS within an organisation:

- The pressures to get applications up and running within the typically short time-scale set for pilots are likely to be such that the emphasis of most pilot studies will be almost entirely technical. Fine intentions about using the pilot to test out assumptions about how a GIS would fit into the organisation are likely to be forgotten in the scramble to get something to show to the boss before the pilot finishes.

- Even as a purely technical exercise, pilot studies are often flawed as they are based upon limited, static, test data sets which do not provide a realistic guide to the performance of the software in a full implementation. Delays which might be imperceptible when the software is being tested upon a small, well-behaved test data set, might well become unbearable when the system is scaled up to full size.

- During a pilot study, the vendor should not be certain of an eventual sale and is still likely to be in pre-sales mode. It may be that levels of support will decline once a sale is secured.

- There is also a danger of premature commitment to a vendor. Although, ostensibly, the purpose of a pilot may be to test the suitability of a particular GIS, in reality an effective commitment is often made as soon as the pilot software is chosen. In most organisations the pilot will be a 'one-shot' event. Given the expense incurred by a GIS pilot, it is unlikely the officers will be given a second chance with another GIS package. Regardless of what it may say in the terms of reference, once a pilot has begun the officers concerned have a common interest with the vendors to achieve a successful outcome.

- Some commentators have been exasperated at the amounts of money which have been spent on repetitious pilot studies. It seems to some external commentators that each local authority and each utility is in danger of insisting upon expensively discovering for itself the already well known. Openshaw (1991) puts this argument vigorously:

  '... the simple bystander might well be forgiven for asking whether the 458 Local Authorities in Britain really do all have unique different and specialised needs that are not found anywhere else. Of course the thought of up to 458 different GIS trials being performed in glorious isolation might well fill the vendors with glee. To others it is a recipe for utter disaster that will eventually backfire on the GIS vendors and suppress the development of GIS in the later part of the 1990s.'

Against these criticisms, it must be recognised that most organisations do believe that they derive benefit from running a pilot study. Perhaps on the sound principle of testing the water before taking the plunge, organisations are unwilling to invest the large amounts of money involved in implementing a corporate GIS without having gained first-hand experience of the realities of the target software and hardware on a small scale. Possibly the adage about the only way to really understand something is to do it yourself is relevant here. It may be frustrating for external observers but perhaps each organisation really does need to run its own pilot study. As Huxhold (1991) puts it, pilot projects are designed to answer the critical question:

> 'Yes, GIS looks like it will work and can improve (performance) but will it work here?'

Implementation

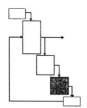 By the end of the Detailed Analysis and Design stage, a project team will know in great detail the nature of the GIS which is required. The next task is to ensure that their plans come to fruition, so that a successful system can be released into the organisation.

In large measure implementation can be seen as a problem of project management and scheduling. Here we will merely outline some of the major concerns:

Project Management: As with any project, the management team will need to establish an implementation plan, identifying the major stages of implementation and planning to ensure that the resources for each stage will be available when needed. This is the stuff of wall-charts and management techniques such as PERT analyses and Gantt charts.

Database Creation: Database creation is critical within implementation. It is costly and time consuming. Setbacks, false-starts and time overruns during the database creation phase can quickly cause management support to wane. A wise project leader, therefore, will take particular care to establish appropriate and realistic data conversion and integration plans. Within these plans issues such as technology (scanning, digitising, line following, etc.), quality (accuracy, documentation, validation, etc.), staffing (in-house, internal sub-contractors, external agencies, etc.), and scheduling (ensuring that some key products can be produced from a partially complete database) will need to be considered.

The New York *GIS Development Guide* (Becker et al., 1997) makes the salient point that GIS belongs to a category of computer applications which only become useful once a database has been completed. With a wordprocessor or spreadsheet, a user can install the software and use it productively immediately. A GIS, however, is an empty shell until an appropriate database is assembled. In view of the difficulties which experience shows often arise during the database creation phase, the New York *Guide* makes database creation the central theme of its approach to GIS implementation. They regard GIS primarily as a database creation exercise.

On reflection, it is a curious feature of GIS implementation that so much attention focuses upon the choice of software and hardware which will be used, when given the pace of technological change, both these features in any project will change within a relatively short period. Even if a project stays with its original suppliers, constant upgrades will mean that the technology upon which a project is initially based will be replaced within a couple of years. In contrast, decisions taken during the data conversion process are likely to have long-term implications for the host organisation.

Application Development: Most GIS projects are based upon proprietary software packages, so that developing 'applications' is most likely to take the form of customising existing software rather than writing extensive new core code. Typically application developers employed by the software vendor will develop applications using specifications jointly agreed with the host organisation's staff.

Roll Out: At some point the GIS system has to be passed from the project team to the users. Clearly this is a critical period in the life of a system. At this point the system begins to impact upon the everyday life of the organisation, and the political and human realities which ultimately will determine the system's success begin to have an effect. If the integration of a new system into an organisation is done insensitively, the system's prospects can be irreparably harmed.

Prior to roll out, the GIS project will have been in the hands of staff who to some extent identify themselves and their careers with the success of the system. After it is launched into the wider organisation, however, the GIS will succeed, or not, upon the degree to which it is adopted as a useful tool by employees who have no particular reason to be interested in GIS per se. The simple presence of GIS software within a company does not mean in any real sense that the company has 'adopted' GIS: only when the company's staff integrate GIS into the daily life of the organisation is it truly 'adopted. The proof of the pudding...

In order to increase the ease with which the GIS is introduced into the wider organisation, the project team will need to have given considerable thought to prior user education, user documentation, user training and user support. A particular concern will be the method of roll out which is employed. Should the GIS team go for a dramatic *'big-bang'* implementation, where the old, probably manual system, is discontinued and the GIS completely takes over, or perhaps there should be a period of *'parallel running'* until users have confidence in the new system? A further alternative would be to have a *'phased run in'* where parts of the new system are introduced in turn. A survey of GIS projects in the UK found evidence of each of these strategies (Medyckyj-Scott and Cornelius, 1993). A 'big-bang' approach was adopted by 31% of the projects, 29% had used parallel running, whilst the remainder had used a phased or evolutionary approach. Which approach works best will vary from organisation to organisation and from project to project (Eason, 1993).

Maintenance and Review

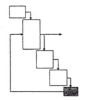 An Information System is never stable. Hardware and software rapidly become outdated and replacements have to be evaluated and justified. User Requirements change. Flaws in the initial design become apparent. The systems development process, therefore, rather like painting the Forth Bridge, never finishes. There is a continuing need to re-evaluate and revise.

Periodically, there is a need to carry out formal audits of the status of the system. Audits provide an opportunity to go back to original documents to compare initial estimates of costs, benefits and time scales with actual outcomes. Audits also provide an opportunity formally to recognise the unexpected outcomes that were not anticipated in the original proposals. Audits can look forward and rechart a system's development to meet changing organisational needs.

Audit documents can be used to argue for continued support, to justify additional spending, to identify problems or opportunities, and to propose corresponding actions. In order that audit reports should be seen to be more than a propaganda exercise on behalf of those involved with the system, audits are often done at 'arm's-length' by independent teams of consultants.

CRITIQUE OF THE METHODOLOGY

As we mentioned in the introduction to this chapter, GIS implementers so far appear to have paid scant attention to the development methodologies which have developed recently in mainstream computing. Indeed few of the published accounts mention the explicit adoption of any named methodology. A fairly pragmatic, common sense, approach seems to have dominated with much mention of 'cost–benefit studies', 'user needs studies' and 'pilot studies' but little sense of there being an over-arching conceptual framework within which these activities take place.

If the often ad hoc methodologies used in GIS developments have drawn upon an underlying conception of how Information Systems should be built then this 'ancestral model' would be the traditional 'life-cycle' model which was developed in mainstream computing during the late 1960s. In the UK the National Computing Centre (NCC) took a lead in developing an approach which broke the development of computer systems down into a series of consecutive stages (Figure 3.6). This approach spread widely during the 1970s being variously referred to as 'conventional systems analysis', 'traditional systems analysis', 'the hard systems approach', 'the systems development life-cycle' or the 'waterfall model'.

The 'typical' GIS methodology described here, and the specific GIS methodologies described in the early GIS texts, are seemingly descendants of this waterfall approach.

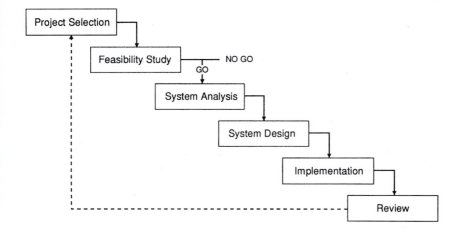

Figure 3.6 Traditional systems analysis/'Waterfall' methodology.

Unfortunately, the traditional life-cycle approach has been heavily criticised in mainstream computing and considerable effort has been devoted to devising new methodologies. Most of the difficulties with Information Systems which have been discussed in previous chapters have emerged from projects which will have been developed using some variant of the traditional life-cycle model, and it hardly seems logical to persevere with a methodology which during the 1970s and 1980s repeatedly produced unsatisfactory results.

Detailed critique of the limitations of the traditional life-cycle approach can be found in a number of texts (e.g. Avison and Fitzgerald, 1988; Eason, 1988; Flynn, 1992). For our purposes, we focus upon four issues:

Technical Emphasis: The traditional life-cycle approach perpetuates the view that developing a computer system is primarily a technical matter. The life-cycle model provides no explicit guidance about understanding the nature of the organisations within which the systems are to be embedded. Users are asked during the early stages of the life-cycle to specify their needs but, thereafter, they are largely excluded from anything more than token involvement until the system is delivered. The traditional methodology assumes that Information Systems development will be led by the technical specialists. Avison and Fitzgerald (1988) condemn this as *pseudo-participation*, saying that user involvement should mean much more than agreeing to be interviewed by an analyst and working extra hours as the operational date for the new system approaches. True participation, they argue, would remove the distinction between 'expert' and 'non-expert', with users and developers working together as a team. In a true participatory methodology, computer specialists become facilitators, rather than experts, helping users to realise the users' ambitions.

GIS User Group Involvement?

'At the outset of the project, a User Group was established. Its remit was to provide objective advice to the Project Team on detailed user requirements and specifications.

The User Group was a major disappointment in the project. Its establishment was well reasoned and theoretically significant. In reality, logistical problems and the demonic insistence of the Project Team to achieve technical results diminished the potential the User Group had.'

Source: Diploma student.

The extent to which GIS implementations have adopted a technical focus is highlighted in a survey by Medyckyj-Scott and Cornelius (reported in Grimshaw, 1994) which shows that some 27% of their respondents reported that 'they did not engage in staff consultations, hold meetings on organisational implications, arrange awareness seminars, have meetings on job impacts or statements of human and organisational requirements'.

Implicitly, the view seems to be taken that the inherent superiority of the new information technologies over the old manual methods is so self-evident that organisational acceptance can be assumed. Implementation therefore is regarded as primarily choosing and fine-tuning the technology (Campbell and Masser, 1995).

Building the Wrong System: Not surprisingly, pseudo-participation can lead to the problems which Bell and Wood-Harper (1992) refer to as stemming from *expert imposition.* Having had insufficient involvement in planning the system, users are left feeling that they have had imposed upon them the system which the 'experts' think they should have, rather than the one they actually wanted. The familiar 'swing' diagram is often used to illustrate this point (Figure 3.7). Campbell and Masser (1995) provide startling evidence of the unhappy effects of 'expert imposition' in GIS from their survey of UK local governments:

> 'Users in several of the case studies commented that they had never wanted a GIS. In these cases it was felt that the technology had been imposed from above to provide information that was of only marginal value to their own work or that of their department. It was evident that the lack of user involvement in the initial decision had caused resentment which was subsequently difficult to overcome.'

Flynn (1992) makes the point very strongly that because of its relative neglect of users there is a real danger that traditional methodology will lead to the wrong systems being developed. He points out that with traditionally developed systems, on average only 8% of the total development budget is spent on determining User Requirements despite the fact that mistakes made during the User Requirements stage are much the most expensive to put right.

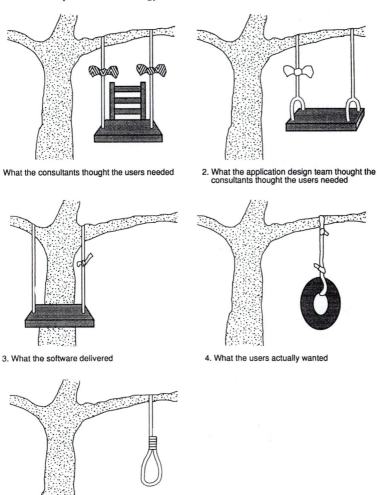

What the consultants thought the users needed

2. What the application design team thought the consultants thought the users needed

3. What the software delivered

4. What the users actually wanted

5. What the users did with it

Figure 3.7 Users don't always get what they wanted.

Stability of Requirements: The life-cycle model assumes that users will be able to provide a definitive set of requirements, and that once stated these requirements will remain stable. There is little scope for second thoughts. In some circumstances these are reasonable assumptions. In the 1960s, when the traditional life-cycle approach was developed, the emphasis in computing was upon developing data-processing systems that often meant automating existing manual systems. In these circumstances it is not unreasonable to assume that users would have a clear view of what was required and that these requirements would remain stable. In the 1980s and 1990s, however, as attempts are made to move computer systems into the higher levels of the organisational triangle, it becomes less and less realistic to assume that users will have a definitive view of their requirements, or that their requirements will be stable over time.

With regard specifically to GIS, there may well also be the difficulty that initially potential users will only have a limited perception of what a GIS might be able to do for them and that the scope of their 'user-requirements' will expand during the development process as they gain a fuller appreciation of the capabilities of the software (Kendrick and Batty, no date).

Neglect of 'Human' Issues: Because of the technological bias of the traditional systems analysis model, there is a relative neglect of the important human issues which will be critical to the success of any Information System. Think of the issues which this chapter has *not* touched upon. We have said nothing about *job content* issues, but clearly any new system is going to alter job contents, raising issues of de-skilling or re-skilling, redundancies, and/or new career progression opportunities. We have said nothing about changes in *operational procedures* which any new Information System will impose. At the very least a new computer system is likely to impose more formal and disciplined work patterns on employees. Often entire work sequences may have to be re-organised. Nothing has been said about the shifts in *power, status and influence* which introducing a computer into an organisation will provoke. Nothing has been said about *health and safety* or *trade union* involvement.

It may well be that in practice these issues would be addressed at some stage in any large project, but it would be better if the methodology gave explicit recognition to them. If introducing Information Systems into organisations is a 'socio-technical' problem, then the development methodology should give as much weight to designing the human half of the system as the technical half.

CONCLUSION

In summary, the traditional life-cycle approach to developing computer systems is an example of *'end-state'* or *'blue print'* planning. It is assumed that the goals which the system is intended to fulfil are stable and uncontroversial; that the boundaries of the project are clearly defined; and that the environment within which the project is taking place is supportive. Once the goals of the system have been fixed, the primary purpose of the development process is to move as efficiently as possible towards the required end state. Thus in the traditional life-cycle approach, once the user requirements have been agreed, the emphasis is upon producing the required technical solution. The stages of the life-cycle are essentially control mechanisms intended to keep projects on track towards their specified goals.

It might be argued that some GIS development projects show the characteristics that would make a traditional life-cycle development methodology appropriate. Perhaps in some operational areas there may be a clear consensus that a map-based procedure should be automated via a GIS system. Perhaps a clear, uncontroversial, well-delineated set of requirements can be agreed and long-term, high-level support can be assured. In these circumstances, the development process will be primarily technical and the traditional life-cycle approach will be appropriate.

In many, perhaps a majority of, GIS developments, however, such circumstances will not apply. Users might be unsure what a GIS might be able to do for them; managers may be unsure whether a GIS in any form is needed; different departments might have differing views upon what sort of GIS development is required. In these circumstances a development methodology which focuses primarily upon the technical issues of the development process seems singularly inappropriate.

In the next two chapters we will look briefly at some current alternatives to the traditional life-cycle model of system development.

SELF-CHECK QUESTIONS

To test your understanding of the material in Chapter Three, you should try to answer the following questions:

1. Attempt to define what is meant by an Information Systems Development Methodology.
2. Within an ISDM what is a 'technique' and a 'tool'?
3. Why do ISDMs need to be based on a philosophy?
4. Why do most ISDMs use graphical techniques?
5. Without referring back to the text, try to draw an annotated diagram to summarise the composite GIS development methodology.
6. If you were in charge of a GIS Working Party which external organisations would you contact and what questions would you ask them?
7. Imagine you are in charge of a User Needs Study to determine whether your organisation needs a GIS:
    — what sorts of people would you want on your interviewing team?
    — which departments/work groups would you want to interview?
    — try to draft out the topic headings you would expect your questionnaire to cover.
8. What are the problems with interpreting User Needs Study reports?
9. Try to write down the categories of COSTS associated with GIS projects.
10. Try to write down the categories of BENEFITS associated with GIS.
11. Why might Cost–Benefit Analysis be more appropriate for evaluating operational activities than for policy level activities?
12. Try to make a list of the advantages and disadvantages of using a Pilot Study as part of a GIS development methodology.
13. Try to list the advantages and disadvantages of Benchmarking.
14. We have argued that many GIS developments seem to have been based on a traditional 'life-cycle/systems analysis' model of development. What are the strengths and weaknesses of this model? When might the conventional model be appropriate? In what circumstances would it be inappropriate?

WHAT YOU HAVE LEARNT IN THIS CHAPTER

* *the meaning of 'methodology' in terms of computer systems development*
  — philosophy, techniques, tools, stages, graphics
* *a composite 'waterfall' GIS development methodology:*
  — Initial Awareness, Building the Business Case, Detailed Design and Analysis, Maintenance and Audit
* *the limitations of the conventional 'waterfall' approach to systems design*

GOING FURTHER

The New York State *GIS Development Guide* provides an excellent companion text for this chapter, describing many of the topics touched upon here in depth. Anyone who is charged with introducing GIS into a large organisation would profit from reading this text, which is available on the WWW via the NCGIA's Core Curriculum project:

BECKER, P., CALKINS, H., COTE, C., FINNERAN, C., HAYES, G. and MURDOCK T., 1997, *GIS Development Guide*, (Albany, New York: Local Government Technology Services, State Archives and Records Administration), (http://www.ncgia.ucsb.edu/education/curricula/giscc/units/u136/).

Several of the early GIS textbooks contain chapters on the GIS development process. The following books, in particular, contain worthwhile material:

ANTENUCCI, J.C., BROWN, K., CROSWELL, P.L. and KEVANY M.J., 1991, *Geographic Information Systems: A Guide to the Technology*, (New York: Van Nostrand Reinhold). (Chapter 10: Implementation. Chapter 4 also contains an extended discussion of Cost–Benefit Analysis in GIS).
ARONOFF, S., 1989, *Geographic Information Systems: A Management Perspective*, (Ottawa: DWL Publications). (Chapter 8: Implementing a GIS).
HUXHOLD, W.E., 1991, *An Introduction to Urban Geographic Information Systems*, (New York: Oxford University Press). (Chapter 7: The Model Urban GIS Project).

As interest in the human and organisational aspects of GIS has grown in recent years a number of 'second wave' GIS texts have been published which make these issues their major focus. These include:

CAMPBELL, H. and MASSER, I., 1995, *GIS and Organisations: How Effective Are GIS in Practice*, (London: Taylor and Francis).
GRIMSHAW, D.J., 1995, *Bringing Geographical Information Systems into Business*, (Harlow: Longman).
HUXHOLD, W.E. and LEVINSOHN, A.G., 1995, *Managing Geographic Information System Projects*, (New York: Oxford University Press).
OBERMEYER, N.J. and PINTO, J.K., 1994, *Managing Geographic Information Systems*, (New York: Guilford Press).

Anyone who wishes to pursue further an analysis of specifically GIS development methodologies would profit by reading Ferrari and Onsrud's extended review of the literature:

FERRARI, R. and ONSRUD, H.J., 1995, *Understanding Guidance on GIS Implementation: A Comprehensive Review*, Technical Report 95-13, NCGIA, Department of Spatial Science and Engineering, (University of Maine: Oraono).

The computer science literature on Information Systems Development Methodologies is understandably vast. The following texts, however, provide good critiques of the conventional waterfall model and were used as sources for this chapter:

AVISON, D.E. and FITZGERALD, G., 1988, *Information Systems Developments: Methodologies, Techniques and Tools*, (Oxford: Blackwell Scientific Publications).
FLYNN, D.J., 1992, *Information Systems Requirements: Determination and Analysis*, (London: McGraw-Hill Book Company).

If you are particularly interested in the economic approach to justifying Information Systems, then the following provides an extended treatment of issues such as cost–benefit analysis:

PARKER, M.M. and BENSON, R.J., 1988, *Information Economics: Linking Business Performance to Information Technology*, (London: Prentice-Hall).

For those interested in reading of the challenges imposed by using cost–benefit techniques specifically with regard to GIS projects, the *International Journal of GIS* has published a number of relevant papers, including:

DICKINSON, H.J. and CALKINS, H.W., 1988, The economic evaluation of implementing a GIS. *International Journal of GIS*, **2**, 4, pp. 307–329.
DICKINSON, H.J. and CALKINS, H.W., 1990, Comments on 'Concerning the economic evaluation of implementing a GIS'. *International Journal of GIS*, **4**, 2, pp. 211–213.
SMITH, D.A. and TOMLINSON, R.F., 1992, Assessing costs and benefits of Geographical Information Systems: Methodologies and implementation issues. *International Journal of GIS*, **3**, 6, pp. 247–256.
WILCOX, D. L., 1990, Concerning the economic evaluation of implementing a GIS. *International Journal of GIS*, **4**, 2, pp. 203–210.
WORRALL, L., 1994, Justifying investment in GIS: A local government perspective. *International Journal of GIS*, **8**, 6, pp. 545–566.

Chapter Four: Alternative Information Systems Development Methodologies: Socio-technic Methodologies

In this chapter we look at some alternative approaches to information system development which have evolved to compensate for the limitations of the traditional systems analysis methodologies. We consider:

* *three 'drivers' which have underpinned many of the newer methodologies: the idea that people matter, the changing nature of business and the enabling power of technology*
* *the evolution of a methodology in the classic tradition under the influence of these drivers*
* *two well-known and widely used socio-technic methodologies and their relation to the frameworks of ideas which underpin them*
* *the idea of participation in system design*
* *the idea of reconciling of socio- and techno-centric views*
* *the appropriateness of each methodology for GIS development*

'A methodology is an explicit way of structuring one's thinking and actions. Methodologies contain model(s) and reflect particular perspectives of "reality" based on a set of particular philosophical paradigms. A methodology should tell you "what" steps to take and "how" to perform those steps but most importantly the reasons "why" those steps should be taken in that particular order.'
Jayaratna, 1994.

'There are four classes of monarch who rule their countries by different methods. The top-class ruler applies silent instructions, leaving his subjects to do what they like and lead their own lives. The second class monarch uses moral principles to influence his subjects and to rule them with benevolence and rewards. The third class ruler controls his subjects with political power and threatens them with punishment. The fourth class monarch makes a show of force and deceives his subjects with stratagems.

The best way of administration is to rule effortlessly without striving. Let the populace do what they like and lead the way of life they wish, thereby bringing the greatest benefit so much so that the populace do not realise their government's achievement, alleging it comes about spontaneously.'
Lao Zi.

INTRODUCTION — SYSTEM DEVELOPMENT DRIVERS

Chapters Two and Three introduced us to the idea that to understand the role of GIS in organisations we had to appreciate first the general role of IS in organisations and secondly that introducing an IS involves some methodology. The traditional systems analysis approach, or the waterfall model, has been the de facto and default methodology used in most IS implementations. In the critical analysis of the model in Chapter Three, we emphasised its shortcomings and the key issues which many IS professionals regard as underlying the failure of systems.

In this chapter we focus on alternative methodologies and give a flavour of the rich response to the problems of the waterfall model. There have been many different responses that attempt to address one or more of the problems listed at the end of Chapter Three. Before we illustrate these methodologies we should, however, try to recognise what the broader picture looks like, because realising the need for new methodologies has not been a simple mechanistic response to the problems. We need to understand something of the context of the change that has driven the creation of new ISDMs.

The motivations which have driven the emergence of the new methodologies can be summarised under three headings:

People Matter: In the last few decades there has developed a philosophy in business which says that people matter. It is not universal, of course, and there are frequent examples in the tabloid press of dictatorial bosses and macho managers who have introduced unpopular working practices or dismissed staff. The fact that such actions make the press is, of course, testimony to the accepted view of what is normal in business relations. What is more normal nowadays is the belief in consultation and participation. The people in an organisation are regarded as valued resources within organisations rather than mere employees who perform tasks under management control. These issues are dealt with more fully in Chapter Six where we consider 'peopleware' in GIS.

Parallel to the changes in attitude to the importance of people in organisations has been an increasing degree to which people are on the ISDM agenda. The fact that the concerns of people had not been fully incorporated into traditional ISD methodologies has been well analysed. Here we see how new methodologies have attempted to incorporate the concerns of people into system design though how this has been done differs considerably between methodologies.

Business is Faster: Business is becoming ever faster. Customers demand prompt delivery. Employees are expected to achieve higher productivity. Product Development Cycles are being shortened. Capital circles the world at electronic speed. Information is instant and all pervasive.

> 'No company is safe.... There is no such thing as a "solid" or even substantial lead over competitors. Too much is changing for any to be complacent. Moreover, the "champ" to "chump" cycles are growing ever shorter.'
> Peters, 1987.

Information is, of course, a major key to the acceleration of business. Technical revolutions in Information Systems during the past three decades have meant that business adjustments can be made faster than before. All this means that introducing information systems has to be made faster and more focused. If the development time of an IS is two years yet an organisation needs to adjust its operation on an annual or six month basis then the IS can never be successful. Development methodologies have had to change in the face of this reality. Today the emphasis in systems development is on *rapid* delivery.

Technology Enables: A third driver that has shaped the evolution of ISDMs is technology. The astounding increase in the power of computers and the parallel developments in software are now part of our culture. Desktop PCs of the present time have the power of mainframes of a decade ago and *the trend continues at the same rate*. The days of the computer users having to write code and issue commands from a keyboard have gone. The graphical display with fourth generation languages and development toolkits are replacing the art of coding and of compiling and debugging. Development times for software projects are a small fraction of a decade or two ago. Changes to software and to systems can be done 'on the fly'. These capabilities have revolutionised the IT industry and the IS in business. The ISDMs discussed later in the next chapter, and particularly those which are linked to *Rapid Applications Development (RAD),* illustrate the ways in which changing technologies can both accelerate and fundamentally change the design and implementation of IS.

In particular, acceleration of the development process introduces the realistic possibility of prototyping as a way not only of testing systems but also of developing the design of systems. This has had a profound effect on methodologies since it allows user participation to a much greater degree and we shall explore what this means for the development process in the next chapter.

New technologies have also meant the emergence of what is called *componentware*. That is software made up not of a single structured package of software routines with a single proprietary interface but of independent packages coupled together through an operating system, such as Windows. This means that software components are embedded in each other so that particular functions, such as GIS, are called and used within other software. So if you want to put a map in your report you call the GIS tool at the appropriate point and embed a map in a document with the full functionality of a GIS not merely as an image. Again the development of components, being based on re-usable software packets, is revolutionising the development of systems. So much so that the methodologies which are appropriate to such situations can be radically different from the traditional waterfall model outlined above.

THE VARIETY OF DEVELOPMENT METHODOLOGIES

Prior to today's systems development methods there were no accepted approaches or ways of designing and implementing information systems. Many systems were developed in an ad hoc fashion with little regard for business needs or usability. Engineering tradition formed the basis for development and there was no experience of the peculiarities of information systems as opposed to other engineering constructions.

Traditional systems analysis, the waterfall model, was developed in response to this somewhat anarchic situation in an effort to provide more coherent and structured ways of developing information systems. The emphasis was on control. The waterfall model is accordingly a highly structured, sequential and predictable set of stages and techniques. People who used the waterfall model knew what they were doing. But it had its problems and systems failure was too frequent and too costly to ignore the tensions it created.

The waterfall model is by no means a single, unified approach to systems development. In fact there are scores if not hundreds of development methodologies based on the idea of the simple sequence of:

— feasibility study
— system investigation
— system analysis
— system design
— implementation

which makes up the waterfall model (Avison and Fitzgerald, 1988). But in the more recent period since the shortcomings of the waterfall model have been realised there have developed many other methodologies which have attempted to deal with the problems which systems developers have faced. Many of these methodologies have emerged only in recent years and against a background of rapid developments in technology. This means that the whole field of methodology is in a state of flux as developers get to grips with the possibilities opened up by changing technology and at the same time attempt to meet the needs of ever more aware management and workers.

Just as with the waterfall model there is no single methodology which represents a simple response to technological drivers. Nor has one been developed simply in response to the increasing speed of business development, nor indeed in response to the growing awareness of the importance of people and the politics of organisations. There is a melting pot of ideas. We tried to build a picture in Chapter Three that methodologies were to be understood against a complex backdrop of problems, methods, technologies and social and other influences. And we want to retain that perspective.

In order to make sense of the complex array of methodologies that have arisen in recent years we will focus on a set of ideas which represent some of the main objectives of these new methodologies. The ideas are in a sense the shortcomings of the traditional waterfall model. Associated with each idea is a methodology that attempts to address the particular problem. It will be clear in what follows, however, that there are no clear divisions between methodologies and no single idea is associated with one methodology alone. But the linking of the ideas and the methodologies gives us an introduction to a complex area of development.

We cannot claim that the ideas and methodologies we describe in this chapter have yet had much impact on GIS. Indeed, on the contrary, our argument is that they have had depressingly little impact upon GIS, and that a greater awareness of alternative ISDMs might go some way to increasing the success rate of GIS projects.

MODERN SYSTEMS DEVELOPMENT METHODOLOGIES

There have been two main streams in the ideas about modern systems development methodologies. On the one hand there have been ideas about the role of individuals, their position in the organisation and the problems which arise when these and other similar issues are ignored. In this chapter we consider methodologies which have arisen in response to these ideas focusing on two: *ETHICS* and *Multiview.*

ETHICS is built on the idea of *participation* as the way of addressing the problems which arise from the neglect of the human issues in IS development especially in complex environments. The whole point of ETHICS is to involve the users and no methodology is as well articulated or as fully developed to do this. It has been widely applied in the UK, Scandinavia and elsewhere and its designer, Enid Mumford (1995) claims considerable success for it. There are no known cases of GIS being developed following the ETHICS methodology but it has points which make it particularly attractive to GIS development.

Multiview is based on *Soft Systems Analysis* and its focus is to try to reconcile the socio- with the techno-centric viewpoints in the development process. Systems are looked at from different points of view and these views are then weighed and reconciled in an attempt to avoid the pitfalls of a single approach. There are few examples of specific GIS applications but Multiview too has many points which make it attractive for GIS development.

In the next chapter we consider a second major set of methodologies which have evolved from the needs of developers who have powerful new technologies at their disposal and who are not therefore constrained by the structured approaches of the waterfall type of methodology. Two methodologies we will focus on which illustrate the impact of technology are the Organic Life Cycle and Evolutionary Delivery. Both are best understood now as essential building blocks of Rapid Application Development (RAD), which as we shall see is coming to dominate the world of IS.

In fact there are signs in the methodologies which are evolving at present of a convergence of what in Chapter Three we termed socio-technic and techno-centric approaches but more of this later. Before we do this, let us consider the developments in the classic model in order first to identify how these ideas have been interpreted in the context of the classical approach and secondly to provide a sort of control for understanding other methodologies.

SSADM: THE CLASSIC MODEL DEVELOPED

The classic waterfall model has been developed in many forms during the last twenty years or more. Well-known versions are Structured Analysis and System Specification (SASS), (DeMarco, 1979), the Yourdon Method (Yourdon, 1989) and *Structured Systems Analysis and Design Methodology (SSADM)*. Here we will consider in more detail only SSADM, which is a public domain method produced by the UK Government Centre for Information Systems, the Central Computer and Telecommunications Agency (CCTA, 1995). It is a useful method to consider out of all those available because it is very well documented (it is available as British Standard BS7738), is in the public domain and is in a continuous process of development and upgrade. Currently SSADM is in versions 4+. SSADM like SASS and Yourdon (and many others) brings together a wide range of techniques for analysis and design and provides a coherent and widely used framework for their use. Only analysis and design are covered by SSADM but it has links to project management methods and quality assurance methods. The project management method PRINCE is also a product of CCTA and the agency has produced guidelines on linking the two.

What we will focus on here are the philosophy and structure of SSADM as an illustration of the classic approach and of the changes being experienced under the influences of the various drivers identified above.

Philosophy of SSADM

> 'SSADM aims to help the IT project team to accurately analyse the requirements for an IT system to support an organisations IT strategy and to design and specify an IT system to cost effectively meet that requirement.'
> Tudor and Tudor, 1996.

The overarching idea behind SSADM and other similar methodologies is a high level of control. Control is achieved through documentation and through structure. The emphasis is on a systematic approach to tasks in a sequential manner even though the sequence can, especially in later versions, have elements of cycling.

SSADM (Figure 4.1) is conceptualised as a set of modules which deliver products. Their products are largely documents that are used by the design team. Control of the information products is through what is conceptualised as an information highway, which links the modules (activities), and the project to managers, who are outside the scope of SSADM.

The techniques used in each of the modular stages are indicated in Figure 4.2.

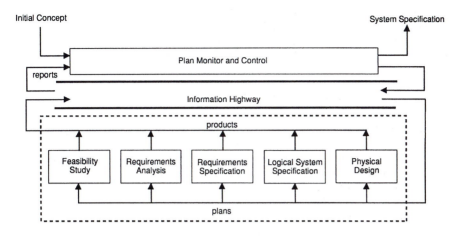

Figure 4.1 SSADM structural model. (Source: Central Computer and Telecommunications Agency, 1996, *SSADM4+ Version 4.3*, London: HMSO.)

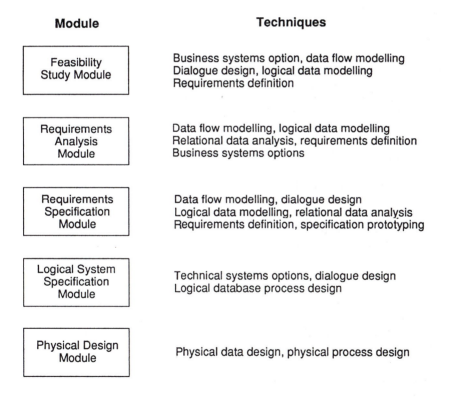

Module	Techniques
Feasibility Study Module	Business systems option, data flow modelling Dialogue design, logical data modelling Requirements definition
Requirements Analysis Module	Data flow modelling, logical data modelling Relational data analysis, requirements definition Business systems options
Requirements Specification Module	Data flow modelling, dialogue design Logical data modelling, relational data analysis Requirements definition, specification prototyping
Logical System Specification Module	Technical systems options, dialogue design Logical database process design
Physical Design Module	Physical data design, physical process design

Figure 4.2 SSADM modules and techniques.

Each of these is an established technical procedure and is widely used in other methods and in information systems modelling. One of the strengths of SSADM is the way in which they are pulled together in a coherent structure. This structure is expressed, at a high level, as a three view model, that is SSADM takes three perspective views of an IS which are used as the basis for cross checking in order to detect and rectify errors. The three views are of:

- **Functions**: Data Flow Modelling yields a processing/data flow view. Function definition builds on this, expressing required system processing in user defined groupings.
- **Events**: Entity/Event Modelling through entity life histories and effect correspondence diagrams provides an event-oriented view.
- **Data**: Logistical Data Modelling provides a data relationship view.

These views are achieved through the various techniques listed in Figure 4.3 and reconciled at various stages throughout the process of analysis and design.

An important part of the philosophy behind SSADM is claimed to be the involvement of users and this is achieved in a number of ways. They are consulted during the fact-finding phases, they are consulted about the products of the various modules and, if management involves users in decisions about the development, they control the proceeding of one phase of the method to the next. This involvement of users in the documented method has increased through the evolution of SSADM and in the latest versions has become more prominent.

Structure of SSADM

The structure of the method is given in Figures 4.1 and 4.2. Each of the five modules involves a set of tasks and outputs (Figure 4.3). The first module, Feasibility Study, may not be necessary depending on the nature of the project. Usually if projects are high risk or politically sensitive such a study should be necessary. However, as IS strategies and business strategies are increasingly becoming aligned and as, increasingly, IS projects are developments from existing systems, then the feasibility study stage is becoming redundant. Projects can begin instead with an initiation document as an input to the Requirements Analysis Module.

Requirements analysis is concerned with information gathering in order fully to understand the business setting and objectives. At this stage a set of business options are presented to users and sponsors from which one must be selected. This process also includes identifying and resolving, between the project team, managers and users, any difficulties which are identified. At this time is begun a Requirements Catalogue which is a central file of requirements for the systems and which is used and added to throughout the project.

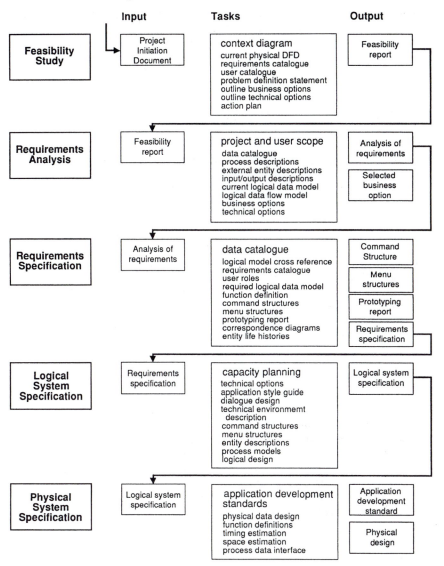

Figure 4.3 SSADM module tasks and outputs.

In addition a User Catalogue is produced giving details of users and their tasks and roles. All these documents form a common point of reference for various stages of the project. Together with the existing data catalogue and the high-level logical data models they form a description of the existing system which, when viewed with the users, forms the outputs of the first module that is expressed in terms of a number of business options.

The Requirement Specification Module takes the project through the next stage to produce the set of products that define the system which has been selected from the Business Systems Options. This is the stage at which detail is added to the data, functions and events which define the system and which will form the basis for a contractual arrangement for the supply of the actual system. This involves revising or updating of the requirements and system objectives so that the final requirement specification is essentially 'a sign off' document for the users 'acceptance of the future system' (Tudor and Tudor, 1995).

The subsequent module on Logical System Specification is concerned with *how* the system will meet the requirements as opposed to *what* the system will do. It considers, amongst other things, how the hardware and software options impact upon users and human resource issues as well as the cost-effectiveness of a system. Its output consists of a logical system specification, in terms of things such as entity, process and interface models on which the physical design can be made.

The Physical Design Model is:

> 'To specify the physical data, process, inputs and outputs using
> the language and features of the chosen physical environment,
> incorporating installation standards.'
> CCTA, 1995.

This involves creating a first cut design 'using a number of guidelines presented in the SSADM Manual, and based on experience and good practice' (Tudor and Tudor, 1995). Optimisation of the design is achieved using a cycle of design–test–optimise–review. The product of this module should be sufficient for implementation of a system.

Recent Developments of SSADM

Recent versions of SSADM (termed v.4+) have attempted to add flexibility to the working of the method and to its positioning in the whole development life-cycle. At a conceptual and procedural level there is a shift of emphasis from the structural model to what is called a System Development Template (Figure 4.4).

The idea is to position the methodology and its techniques with their inputs, outputs and procedures into a framework which more systematically and formally links it to the other elements of the business environment. The template pays specific attention to the 'Decision Environment' to users, to policies and procedures and to the system construction process. At the core is the three-part SSADM model of conceptual model, internal design and external design but in this too there is an increasing emphasis on the need to customise and take account of specific project and organisational objectives with Business Activity Models and Work Practice Modelling.

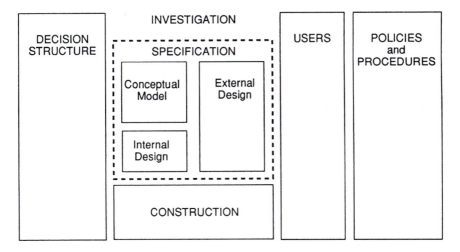

Figure 4.4 SSADM system development template. (Source: Central Computer and Telecommunications Agency, 1996, *SSADM4+ Version 4.3*, London: HMSO.)

An interesting and particular development in the development of the SSADM repertoire is the inclusion of Object-Oriented (O-O) ideas and techniques (CCTA, 1995). As you will see in the next chapter, O-O techniques are an important part of the approach to Rapid Development. As such, the underlying rationale of SSADM and O-O are somewhat at variance. Perhaps it is a testimony to the strength of the SSADM culture that O-O techniques can be incorporated.

These changes are in part a response to the sorts of drivers identified above but also they are to be understood in terms of a move to merge SSADM with EUROMETHOD. This is an approach being developed at the EU level to structuring the supplier–contractor relationship based on modern ideas of the development life-cycle.

SSADM and GIS

SSADM has been extended to make specific reference to GIS. A report published in 1994 (CCTA, 1994), and referring therefore to SSADM v.4, considers how the need to handle the geographic information affects each of the modules and stages of the method.

The main points from CCTA's analysis are:

- Geographic data places specific and significant demands on the design process
- Geographic data entities and procedures are inherently complex
- Users place high demands on the analysis of requirements
- Specific knowledge of geographic data and GIS is essential for good design.

But their overall conclusion is that:

> SSADM is capable, *as it is*, of handling geographic information system analysis and design.

In reality, however, the adoption of SSADM by organisations undertaking GIS projects has been very limited indeed. A survey of almost fifty UK companies (Keenan, 1996) showed that the methodology had been used in some GIS projects but almost always only partially. In almost all cases there were local adaptations or the methodology was followed in name only. A survey of twenty-one Irish GIS projects (O'Donogue, 1997) reported that only six had adopted a formal methodology and, of these, three had claimed to have used SSADM. However, none had used SSADM in full, selecting only the techniques and tools that were considered suitable to the local situation rather than follow the logic of the methodology.

Why SSADM is not being used is unclear. There is insufficient evidence either to demonstrate that the method is intrinsically inappropriate for GIS or that organisations are reluctant to adopt any kind of structured method irrespective of the type of project. Certainly the use of structured methods is not as universal as it may seem. Significantly, one of the missions of CCTA is to promote the adoption of structured methods in IS projects. So the seeming reluctance to use SSADM or any other methodology in GIS projects is perhaps only part of a wider malaise.

Issues of SSADM

SSADM has undergone several substantial revisions and recent versions have incorporated elements derived from other concepts of the development life cycle. It remains, however, a linear and highly structured method which is based on the simple concept of the waterfall model.

If we visit again the list of assumptions of the waterfall model (Figure 4.5) we can ask whether or not they apply to SSADM and its modifications.

> Requirements are known by the users.
> Requirements are stable.
> Requirements can be and are fully specified.
> Technology remains stable.
> Business environment remains stable.
> Developers understand business needs.
> Future system is acceptable to users.
> A system can be designed in one go.

Figure 4.5 Assumptions of the waterfall model.

These assumptions relate to the linear structure and to the level of participation by users. The problems with the pure waterfall model follow from its rigid linear sequence and its lack of user participation. Clearly SSADM retains a highly linear structure. The modifications to it with possibilities of cycling back in the development process are there but the dynamics of the method mitigate against these being radical. As McConnell says of the modified waterfall method (the salmon life-cycle):

'You are allowed to swim upstream but the effort might kill you!'
McConnell, 1996.

Clearly, the methodology is still very much in the hands of the development team. As Jayaratna (1995) would say, it follows a positivist paradigm with control by managers. The claim by Tudor and Tudor (1996) that:

'a fundamental principle underlying SSADM is that the system belongs to the users and hence their participation in the development process is essential'

doesn't quite ring true and there are a number of reasons why user involvement has to be controlled. It occurs at several points in the methodology but in specified ways and, in the context of an IS development method, is politically weak. Users are required to provide certain defined sorts of input and only at specific stages. The eventual users of an information system are allowed to make an input to saying *what* the system should be but not into *how* that system should be developed. This *how* bit is important since it defines the scope of any analysis and in such highly structured methods as SSADM there has to be a focus on formal characteristics of information and ignoring of these informal aspects which are difficult to computerise.

The methodology works through documentation. Documentation is produced by the development team for themselves and for managers. This is the product of the methodology, the means of control and the output to any subsequent stage. The emphasis on documentation works against flexibility in scope or sequence of the methodology and its techniques work against those who are not in the team.

There is a further set of assumptions about SSADM which are not normally examined in the literature but which are clearly important from the descriptions of the methodology. These relate to the skill of developers who are using it.

There are in the methodology a number of stages at which there has to be reconciling of different products or at which users make important inputs — in discussion requirements, in assessing the requirements catalogue, in agreeing functional definitions and so on. At each stage there is a need for reconciling not only abstract structures but also opposing views as well as a need for articulating views. No doubt also there is a need for negotiation. However, the method has no indication on how these needs may be approached.

Similarly, it is clear, even in the method, that users, designers and managers may have different mental constructs not only of the existing IS but of the future one and of the means for going from one to the other. Versions 4+ of SSADM include business activities modelling and work practice modelling and are designed to link to EUROMETHOD which has affinities with soft systems methods. These extensions may address these issues to some extent but the origins of the methodology remain dominant.

It remains to be seen whether or not the realities of everyday projects allow the modifications to SSADM to be effective. The modifications are mostly extensions done in response to developments which we identify as the 'drivers' causing change. These drivers are, however, more effectively expressed in other methods, such as the ones we will explore below. The developments of SSADM are not radical modifications of the core elements of the methodology which remain essentially pure waterfall.

Having spent some time considering SSADM as an example of a methodology with its roots in the classic waterfall which is now attempting to accommodate current concerns within its structure, we will now turn to methodologies which are more firmly grounded in the socio-technical approach.

PARTICIPATION IN SYSTEMS DEVELOPMENT

'The Masses are more knowing and more constant than is a Prince.'
Niccolo Machiavelli, 1519.

ETHICS: Effective Technical and Human Implementation of Computer-based Systems

ETHICS is both an idea and an acronym. It is meant to introduce an ethical basis to systems design. By that is meant a consideration of the values of the people who will be affected by a system. Of course, advocates of such an approach have a fundamentally different emphasis on information systems, organisations and jobs from those who advocate more techno-centric approaches. Personal, group and social agendas are more in focus and financial and technical ones less so than in other approaches.

ETHICS is based on the belief that for a system to be effective the technology must 'fit' closely with the organisation in which it is to be used (Mumford, 1983). The approach focuses on the *changes* involved in introducing technology. As such it has three objectives: the *first* is to establish a position which includes all users of a system in design. The reason is that if people can influence their own work situation then job satisfaction and efficiency are more likely. Satisfaction because users are best able to diagnose their own needs compared to outside specialists. Efficiency because users have the best day-to-day knowledge of information needs and use, and of the myriad of work problems associated with using any system. Participation in design is believed also to create commitment to making a new system work.

The *second* objective is to enable groups involved in systems design to set job satisfaction objectives in addition to technical objectives. The reason for this is that unless satisfaction is made *explicit* the outcomes will be unpredictable because they have not been planned for. The articulation of satisfaction objectives helps ensure human satisfaction with systems.

The *third* objective is to ensure that any new technical system is surrounded by a compatible, functioning organisation system. This recognises that technical design is only part of a complex design process that includes not only the technical system but also what surrounds and interacts with it. This third objective addresses issues of a man–machine interaction, work procedures, roles and responsibilities and management.

Participation is seen as the key to design. It is a way of working with greater knowledge and understanding of a system but perhaps more importantly it serves as a mechanism or expression of *consent*. What ETHICS seeks to define is the *structure* of participation and the *content* of participation and the *process* of participation.

There are different forms and processes of participation, both direct and indirect and one or a combination of types, viz. consultative, democratic and responsible. Participation is seen as scoping a project, including scoping participation by setting boundaries to decision making. It is concerned with knowledge and learning and with power. It is a basis for negotiation and compromise.

Comprehensive participation would involve (Mumford, 1983):

- participation in the initiation of a project — agreeing that it shall go ahead
- diagnosis and specification of existing problems and needs
- setting of business, human and technical objectives — design of alternative solutions
- feasibility study and evaluation of alternatives
- detailed design of human and technical work systems and procedures
- implementation of system
- evaluation of working system.

This sequence might look familiar. It should. It is almost an exact mirror of the traditional waterfall approach or of any of the structured methods with some additions. But, it is a whole culture away because of the premises of the approach and the participatory methods that are employed.

Participation has its problems that are common to all social interaction. There must be trust and so communication is important as well as transparent motives. Working together exposes conflicts of interest and generates stress. And the skills of managing people, working in groups and assuming particular roles are in all probability more difficult than the technical skills of using systems. Like all methodologies ETHICS requires good management by people with special skills of communication and facilitation.

Central to the methodology and what sets it apart from other methodologies is the idea of job satisfaction. This is defined as (Mumford, 1983):

the employees' job expectation

a good 'fit' between

job requirements as defined by the organisation

An analytical framework for describing and measuring job satisfaction deals with five aspects of fit (Parsons and Shils, 1951). They are:

— the knowledge fit
— the psychological fit
— the efficiency fit
— the task structure fit
— the ethical (social value) fit

and good fits exist when:

- employees believe their skills are being adequately used and they are being assisted to develop these skills
- employees believe that personal aspirations are recognised and met
- employees achieve support, rewards and the controls they require
- technology and task structure produce a work situation where employees have the variety and opportunity which fits their personal requirements
- the organisation is able to meet employee values on communication, consultation, participation and other aspects of human relationships.

Designing an information system based on these principles and incorporating the difficult issues of ensuring job satisfaction involves, like all structured methods, techno-centric or socio-technic, a high level of control. This is (has to be) achieved by modularising (breaking down) the work and following a structured sequence of events.   Documentation is essential. Mumford (1983) provides a fifteen-step procedure for ETHICS summarised as:

1   Why change? — this is the first question of the design group. It is the basis for future consensus and participation (only if, presumably, there is a positive answer!).
2   System boundaries — the group identifies the boundaries of the system and where it interfaces with other systems in terms of business activities, existing technology, organisation parts, environment.
3   Description of the existing system — this is to educate the group about how the whole system works.
4   Definition of key objectives.
5   Definition of key tasks.
6   Definition of information needs.
7   Diagnosis of efficiency needs.
8   Diagnosis of job satisfaction needs.
9   Future analysis.
10   Specifying and weighting efficiency and job satisfaction needs and objectives.
11   Organisational design of the new system.
12   Technical options.
13   Preparation of a detailed work design.
14   Implementation.
15   Evaluation.

Critics of ETHICS question the ability of users to design complex technical systems and that of managers to follow and manage the methodology. ETHICS is not easy, but the proponents argue that the benefits from making the effort and adopting the approach and the improvements in relations as well as practice are worth it.

ETHICS has been applied in many organisations and has been successful. By that we mean that effective systems have been designed and are used which have been created using the methodology. Mumford (1995) makes clear also that the acceptance and introduction of the ETHICS approach has a wider impact on organisations than simply improving IT. Using ETHICS involves a whole baggage of management and employee attitudes, norms and practices which are based on co-operation, reasonableness and rationality and these qualities have a much wider relevance and impact than those simply related to the development life-cycle. Indeed Mumford (1995) places considerable emphasis on the evolution of total organisational or business philosophy so that 'all parts are co-ordinated, so moving together in their closely knit and adjusted activities, so linking interlocking, interrelating that they make a working unit' (Mary Parker Folett, quoted in Mumford, 1995). Underlying the whole philosophy are ideas of freedom in work and power *sharing*, *joint* responsibility and *multiple* leadership.

Participation in Systems Development and GIS

The ideas expressed by Mumford in espousing ETHICS strike familiar chords for the field of GIS. Since there are no examples of ETHICS used for GIS in practice let us consider why this might be so from first principles. What GIS have to offer is, through the use of spatially referenced data, a powerful means of structuring data. As Fotheringham (1996) says:

> 'The defining characteristic of spatial data — the thing that makes
> it special — is that it is tied to locations. This means that each
> piece of data has a unique set of relations to all other data.'

GIS add to the capabilities of information systems because of this spatial component. We can argue that therefore adopting a GIS approach to data handling is likely to be significant for business processes. Bestebreurtje (1997) points out also that GIS are characterised by being complex, which is a consequence of their structure and their role in organisations:

> 'GIS projects have kept out of the mainstream of IS projects till
> recently since they are considered to be "different and complex"
> but this is the fate of every new technology. ...Suddenly there are
> multi-departmental or even multi-organisational GIS projects ...but
> there is little or no experience in how to handle such projects. As a
> matter of fact there isn't much experience with other non-GIS
> projects of comparable magnitude and complexity as well.'

This complexity is reflected elsewhere in the view of GIS as being inter-disciplinary and this point is taken up again in the final chapter:

> '...a GIS is a multi-disciplinary approach to information handling which involves a wide range of technical and organisational problems. The same organisation and people problems which have been a feature of other information systems have also been a feature of GIS implementation. However, the unique characteristics of a GIS, in that it is multi-disciplinary (and) a different way of handling and viewing data, (mean that) the organisational changes brought about by its implementation are unpredictable (which) makes its implementation more difficult than other systems.'
> O'Donoghue, 1997.

The immediate, inescapable consequence is that using GIS is likely to affect people and organisations to perhaps a greater degree than most other information systems and in any case the role of GIS within an organisation is likely to be complex. For a multitude of reasons we can see that consent of users is going to be a major factor in the success of a GIS since so many are likely to be affected by it and to a significant degree. It is exactly such situations that ETHICS is designed for.

RECONCILING POINTS OF VIEW

Soft Systems and Multiview

Soft Systems Theory is based on an approach to the world which values people and their wishes above externally defined constructs such as organisations and information systems. Above all it is an approach which is essentially 'interpretative', that is, it doesn't espouse or accept that there is any single view of reality but that individuals or groups can have their own interpretation of reality which can not be argued to be inferior to any other or to opposing views of reality. By extension, in order to understand situations information systems, professionals undertaking some methodology based on Soft Systems Analysis (SSA) need to value and use debate and participation. Such debate is an essential part of the associated methodologies and indeed dominates some such as Multiview and ETHICS.

As an approach to problem solving, the framework provided by soft systems is open-minded and recognises the complexity of real situations. Indeed, as Jayaratna (1994) points out, it is the only framework which recognises the idea of a 'situation', rather than dealing with externally defined groups, roles and processes. The complex mix assumed for real situations is reflected in the approach and methods of the soft systems based methodologies such as Multiview as we shall see below.

The soft systems approach was developed by Peter Checkland (1981) as a methodology by which the problems of complex ***human activity systems***, particularly business systems, could be investigated. Checkland's central argument is that conventional systems analysis, which he terms 'hard' systems analysis, has proved to be an inappropriate vehicle for investigating human systems, and that the valuable insights available from systems theory needed to be recast into a softer, more interpretative methodology for investigating human activity systems.

Systems ideas, and particularly systems analysis, are fundamental to Computer Science. Many computer professionals call themselves 'Systems Analysts', 'Systems Supervisors', 'Systems Operators', etc. Also, as we have already seen, the conventional Information Systems Development Methodology which emerged in the 1960s is sometimes referred to simply as 'systems analysis'.

Computing, of course, is not alone in feeling the impact of systems concepts. Systems ideas have penetrated almost every area of study and, indeed, everyday life. Anyone who is a town planner, geographer, computer scientist, biologist, engineer, economist, sociologist, management scientist, etc., is likely to have been exposed to 'A Systems Approach to ......' at some stage in their careers.

Although systems analysis in computing has conventionally been within what Checkland refers to as the 'hard' systems analysis tradition, several methodologies are now looking to 'soft' systems analysis as a means of understanding the organisational complexities of Information Systems development.

The development of the soft systems approach represents an attempt to complement the hard systems analysis with a more humanistic, interpretative approach. Soft systems enthusiasts do not say that hard analysis is wrong. On the contrary it is recognised that there are 'hard' types of systems and 'hard' problem areas where 'hard' techniques are appropriate. Rather soft systems enthusiasts argue there are also 'soft' types of systems which require 'soft' techniques. In particular systems which involve human beings — e.g. families, cities, organisations, and *information systems* — cannot be fully understood by formal, hard analysis. Attempts to apply hard analysis to human activity systems were misguided and need to be supplanted by a method based upon a more realistic metaphor.

Soft Systems Analysis

Soft Systems Analysis uses systems concepts in an interpretative manner. Checkland's aim is to retain the core concepts of systems thinking — holistic thought, hierarchy, emergence, **entitation**, etc. — but to reject the quantitative analytical approaches which have traditionally been associated with systems analysis. Notions of optimisation and means–ends planning are eschewed in favour of learning or consultative processes of enquiry.

Although soft systems analysis is intended as a set of general purpose problem structuring techniques, it has attracted interest within Computer Science as an appropriate way of exploring the information systems needs of organisations. It is suggested that the soft systems approach will be particularly appropriate where there are doubts within an organisation about whether an information system is needed, and if so what sort of system is required. In these circumstances determining requirements will demand a prolonged period of discussion, debate and bargaining from which an agreed user requirement might eventually emerge, and soft systems analysis is seen as a means of structuring this crucial initial period.

Figure 4.6 provides a key to the methodology. Checkland emphasises that the stages indicated in the figure represent guidelines which will need to be adapted to fit the requirements of each analysis, rather than a rigid template to be imposed on to every problem. Also progression between the stages is likely to be iterative, rather than linear.

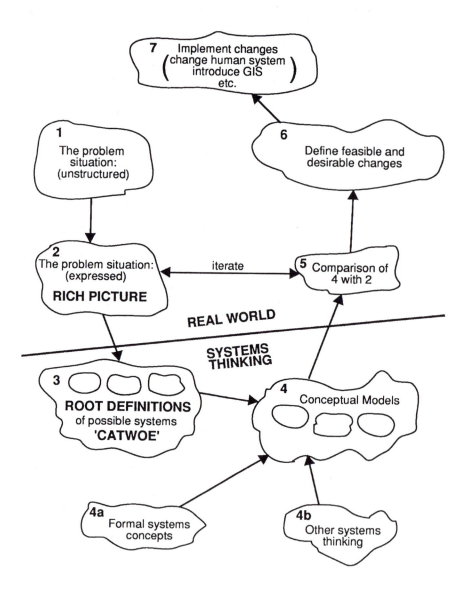

Figure 4.6 Checkland's soft system methodology. (Source: modified from Checkland, P., 1981, *Systems Thinking, Systems Practice*, Chichester: John Wiley and Sons, Figure 6. Used with permission.)

Stages 1 and 2 form the *expression* phase of the methodology. It is in these phases that the analyst attempts to understand the nature of the problem and the organisation within which the problem is located. It cannot be assumed that the nature of the 'problem' will be self-evident. Often analysts are engaged because management within an organisation is aware in general terms that 'something is wrong' or 'something needs doing', and the analyst's first task is to help managers to understand and define their problem more clearly.

In SSA problems are not regarded as external realities having an existence independent of the individuals involved with them. On the contrary, 'problems' are constructs, defined by the perceptions of those affected by them. The first problem for the analyst, therefore, is to achieve an acceptable definition of what the problem shall be deemed to be. Checkland cautions strongly against prematurely defining the nature of the problem. Analysts should be aware of the danger of imposing their own prejudices and past experience upon new situations, and need also to be aware of the complex currents of opinions which can exist within organisations about whether a problem exists. Soft systems analysis, therefore, recommends that analyses start from an assumption that the analyst has been called into a broad, fuzzy, unstructured problem area, and that a considerable initial investigative and consultative effort will be needed before a structured understanding of the problem can emerge. The initial investigations should be open-minded and non-judgmental. The temptation prematurely to classify a problem as belonging to a 'problem type' should be resisted.

Bell and Wood-Harper (1992) provide a list of the techniques that an analyst might use during the expression phases of the methodology, including interviews with stakeholders, regular discussion groups, workshops, and observation techniques. The intention should be to capture all relevant information about a problem: quantitative and qualitative, objective and subjective, official information and gossip.

The distinctive technique of soft systems analysis at this stage of the methodology, however, is the **rich picture** technique. A 'rich picture' is essentially a cartoon (Figure 4.7) in which the analyst attempts to portray her understanding of a problem in diagrammatic and pictorial form. The rich pictures will include both subjective and objective aspects of the problem area. Rich pictures usually contain elements such as the clients of the system, its owners, people working in it, tasks being performed, relationships, areas of conflict and stress — indeed anything which is of relevance to understanding the system. Rich pictures provide a means for the analyst to document her increasing understanding of the problem at hand, and also provide a valuable means of stimulating discussions with clients. Avison and Fitzgerald (1988) suggest that the rich picture technique can draw out parts of the 'iceberg' of a problem which normally remain hidden when using traditional methods of investigation. The example given in Figure 4.7 relates to a situation anonymised from a real situation of a bureau supplying GIS data to users who have to register first through local agents — a situation fraught with problems.

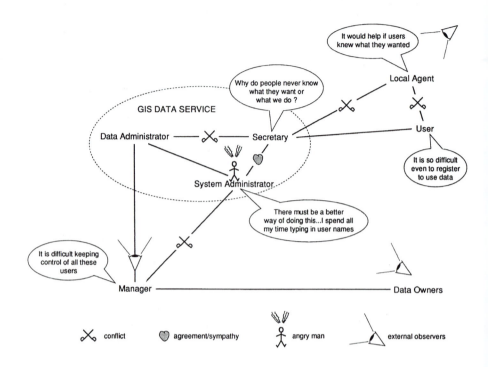

Figure 4.7 An example of a rich picture.

The method by which rich pictures are developed (and indeed the degree of artistic flair with which they are drawn) will vary from analyst to analyst. Naughton (1984), however, provides a checklist of items which might be included in a rich picture diagram:

Structures: These are parts of the situation which are stable or slow to change. Items such as the physical layouts, control hierarchies, job titles, departmental boundaries, etc. might feature here. The structures noted can be both formal (e.g. the GIS working party) and informal (e.g. 'senior staff enthusiastic for GIS').

Processes: These are things which are in a state of change — activities within the structure. Again these can be formal (the systems manager 'reports' to the committee) and informal (the 'systems manager travels to work with the committee secretary'). Often it can seen wrong or impolite to include information about friendships, gossip, antagonisms, etc., in a professional analysis, but if these issues are likely to determine the outcome of a situation they should be recorded.

Interaction between Structures and Processes: Is there any tension between structures and processes? Often structures, being slow to change, lag behind more rapidly changing processes and informal 'work-arounds' emerge to compensate for outmoded structures.

The particular purpose of a rich picture diagram is to capture the types of personal and organisational information which falls outside the scope of more objective, reductionist forms of analysis. From the rich picture there are likely to emerge some major tasks and issues which will become the focus of later stages of the methodology.

Stage Three requires the analyst to identify relevant systems and *root definitions* for these systems. Here we meet one of the key aspects of SSA which is the importance of 'viewpoint' (or *'Weltanschauung'*) within the methodology. In the introduction to systems concepts provided above, we made the point that systems are constructs which exist in the minds of people, rather than real things. Checkland's methodology pursues this idea, by arguing that the analyst will find within an organisation many different possible definitions of a system. Different actors within the problem area will have differing views about what the problem is and what its solution might be — they literally see things differently. The extent of these differences should become apparent during the expression stages. The next stage requires the analyst to explore alternative definitions of the problems and, in discussion with the stakeholders of the problems, to agree, hopefully, a consensus definition which can form the basis for subsequent analysis (or at least agree a limited number of alternative definitions which are to be pursued).

A root definition is essentially a sentence that describes, in abstract, the fundamental nature of a system *when viewed from a particular viewpoint*. As a guide to the construction of root definitions, Checkland provides his *CATWOE* elements, by which he means that a complete root definition should identify the Client (C), the Actors (A), the Transformation (T), the Weltanschauung (W), the Ownership (O) and the Environmental Constraints of the system (E):

C	'Client'	The persons, or groups, who benefit from the outputs of the system.
A	'Actors'	People who are within the system and carryout its functions.
T	'Transformation'	What the system does. The basic purpose of the system.
W	'Weltanschauung'	The viewpoint from which the system is being considered.
O	'Ownership'	Those who have the power to determine how the system performs. Often a larger system of which the target system is a sub-component.
E	'Environmental Constraints'	Other systems with which the target system interacts and which impose constraints or pressures upon the target system.

In less formal terms, the CATWOE elements require that the root definition should state '*who* is doing *what* to *whom,* and to whom they are *answerable*, what *assumptions* are being made and in *what* environment is it happening?' (Avison and Fitzgerald, 1992).

Requiring the analyst to summarise her knowledge of a system into a concise, often single sentence, definition might seem an artificial device, but in practice, this procedure forces the analyst to focus very clearly upon the problem and to ask searching questions about her understanding of it. Producing a set of possible viewpoints and associated root definitions, and asking participants in the problem area to comment upon them can be a very good way of revealing tensions and differences of opinions. To give an example, Avison and Fitzgerald (1988) point out that one root definition for a prison might be 'a system to rehabilitate criminals', another 'a system to train criminals', another 'a system to punish criminals' and yet another 'a storage system for unwanted people'. Each would lead to a dramatically different model for developing the penal system.

At the end of this stage, a decision will need to be taken about which root definition is to form the basis for future development, or perhaps it might be decided to pursue a number of alternative roots. If it proves impossible to get the various stakeholders in a problem situation to agree which root definition(s) should be proceeded with, then the technique has proved its worth as it has highlighted fundamental divisions within the organisation which would make further detailed analysis futile. In our terms, if stakeholders cannot agree what the fundamental purpose of a GIS should be, then there is little point in going further.

Assuming that a root definition(s) can be agreed as a basis for further analysis, the **fourth stage** in the methodology is to construct a ***conceptual model*** showing the actions which would be needed to achieve the transformation specified in the adopted root definition. In soft systems analysis a model is a process diagram showing activities usually in a time sequence. The conceptual model is not supposed to be a stylised description of the existing system, but rather is supposed to be a logical description of the minimum set of processes which are required to achieve the transformation specified in the root definition. The conceptual model, therefore, is an abstract statement from which all references to specific personalities have been removed.

Bell and Wood-Harper (1992) provide a checklist by which a conceptual model can be developed:

- Review the consensus root definition to form an impression of the type of system that will be necessary to carry out the transformation that has been agreed.
- Put together a list of the *verbs* that will be required to achieve the required transformation.
- Thinking in terms of a simple system (input, process, output), use *nouns* to describe the inputs and outputs to and from the activities described by the verbs.
- Organise the activities (verbs) in a logical, time sequential, order and use arrows to join the activities together. The arrows symbolise information or energy or material or some other form of dependency.

A simple example of a conceptual model relating to the rich picture of Figure 4.7, is given in Figure 4.8. For complex systems it may be desirable to produce a series of model diagrams at different levels to capture emergent properties at appropriate points in the hierarchy of possible systems.

Stage Five takes the methodology back into the 'real world' of the organisation and requires the analyst to compare the conceptual model of the proposed system with aspects of the rich picture developed in Stage Two in the presence of clients. The purpose of this stage is to stimulate discussion in order to improve the conceptual model. A number of techniques can be used to stimulate such discussions. For example, participants can be asked to imagine that the conceptual model has been implemented and to work through likely events. Alternatively, the analyst could draw up a model of the actual system and 'overlay' it with the logical conceptual model to highlight areas of change. Often questioning the conceptual model will require a redesign of the rich picture and a further iteration of the methodology.

Stage Six assumes that from the comparison of the conceptual model with the actual situation as represented in the rich picture a series of recommendations for change will emerge which will then need to be considered for action. During Stage Six participants discuss which of the agenda of possible changes are desirable and culturally feasible. Sometimes there will be items on the agenda which are undoubtedly desirable in the sense that they are technically achievable and would lead to improved performance, but which are unacceptable to the culture of the personnel within the situation. In the spirit of soft systems analysis the analyst should accept the verdict of the actors. To try to force a solution upon unwilling actors is likely to lead to expensively failed projects. The whole thrust of SSA is that changes will not happen unless those directly involved with them are persuaded of their merits.

Stage Seven implements any changes that are approved in Stage Six.

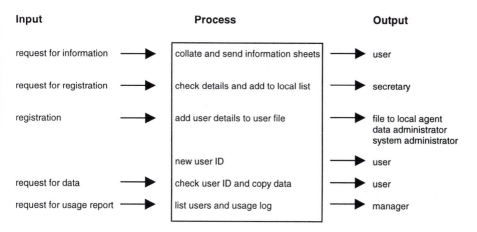

Figure 4.8 An example of a conceptual model.

Multiview

A starting point of the Multiview ISDM is that any information systems project should be considered from at least five different perspectives (see Figure 4.9). Each ring in the diagram represents the concerns of a different group of participants in the process and asks a different question.

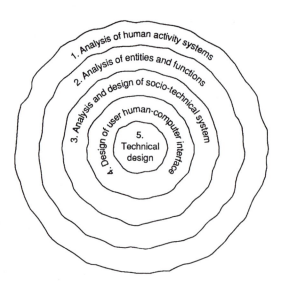

Figure 4.9 Overview of Multiview. (Source: Avison and Wood-Harper, 1990, *Multiview, An Exploration in Information Systems Development*, Oxford: Blackwell Scientific Publications, Figure 2.4. Used with permission.)

Computer Scientists have traditionally focused their interests on the inner ring (issue 5) of the diagram. Many of the early Information Systems Development Methodologies focused primarily upon the area covered by the second (issue 4) ring. The argument behind Multiview, however, is that the success of any information system will depend at least as much upon the issues covered in the other rings. Multiview insists that these broader issues shall be formally considered in the methodology and that all interested parties shall have a voice in the development process. Senior management will be concerned with issues associated with the outer ring. Trades unionists and health and safety officers will be interested in the next ring. Users will have a major concern with issues in the third ring. Figure 4.10 provides an outline of the stages of the Multiview approach.

Briefly the stages are:

1. Analysis of Human Activity System: Here Checkland's soft systems methodology is used to help the analyst to understand the nature of the problem. Multiview does not assume that a computerised information system will always be required. As a result of a SSM analysis an improved Human Activity System might result which resolves the problem. In such cases the methodology stops here.

2. Analysis of Entities and Functions: Assuming a computerised information system is recommended from stage one, the next stage is to develop a model of the entities and functions which will be required by the system. The starting point for this stage is the root definition (and conceptual model) produced in stage one, the root definition being progressively decomposed into necessary sub-functions. This stage of Multiview is similar to the data analysis and modelling stages found in more conventional methodologies.

3. Analysis and Design of the Socio-technical System: Here Multiview adopts an approach similar to that used by ETHICS, in that it emphasises the need to design an Information System which provides a 'good fit' with the needs of the people who will form part of the system. Considerations such as morale and job satisfaction are just as important to Multiview as are technical issues.

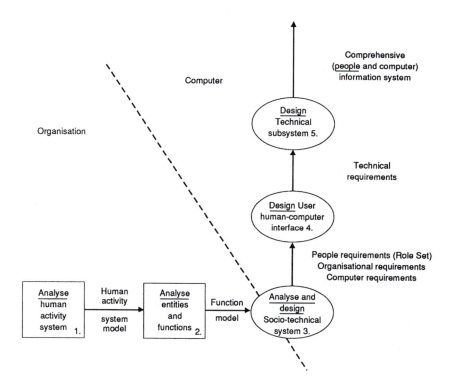

Figure 4.10 Stages of the Multiview methodology. (Source : modified from Avison and Fitzgerald, 1988, *Information Systems Development : Methodologies, Techniques and Tools*, Oxford: Blackwell Science, Figure 6.29. Used with permission.)

The approach Multiview uses in order to balance social considerations against technical ones focuses upon trading off social alternatives against technical ones (Figure 4.11). Social objectives are identified (e.g. provide job satisfaction, safeguard employment, be relatively self-sufficient, improve the professional status of the department), as are technical objectives (e.g. inform management, improve timeliness, increase information processing capacity). For each social and technical objective a range of alternative solutions can be envisaged, e.g. for social objectives — increase staff numbers, provide in-house training, use agency staff: for technical objectives — enhance manual systems, purchase new systems, introduce networked micros, purchase a mini, etc. These social and technical alternatives are then cross-matched and evaluated using ranking systems that consider cost and constraints until a best balance between technical and social alternatives is achieved.

The product of this stage of Multiview is the identification of the set of necessary 'computer tasks' and 'people tasks' that will be required by the chosen solution.

4. Design of Human-computer Interface: The methodology here focuses upon the technical design of the *human-computer interface*. In line with Multiview's philosophy that Information Systems to be successful must be sympathetic to the needs of the people who will be expected to use them, considerable emphasis is placed upon designing a computer interface which is appropriate to the skill levels of users. Should the interface be command-driven, menus driven, form driven or use a mouse/windows interface? How should help screens be designed? Should artificial intelligence be used to prompt users? Wood-Harper and his colleagues emphasise the role of prototyping in order to achieve an appropriate interface.

5. Design of the Technical Subsystem: The inputs to this stage are the outputs from Stage 2 (entity and process models) and Stage 4 (interface requirements). Social considerations have already been explicitly considered in earlier stages of the methodology and so such considerations will be embedded within the specifications of the system which are input to this stage. The final stage of the methodology is thus primarily a technical exercise to develop a computer subsystem that will efficiently meet the requirements.

In summary, Multiview is a hybrid approach which attempts to use appropriate techniques at different stages of the development process. When dealing with the 'soft' aspects of development — trying to define requirements, dealing with social issues — Multiview borrows soft techniques from soft systems methodology and ETHICS (Mumford, 1985). When dealing with 'hard' aspects — information modelling and developing technical specifications — Multiview reverts to more conventional techniques similar to those found in traditional ISDMs.

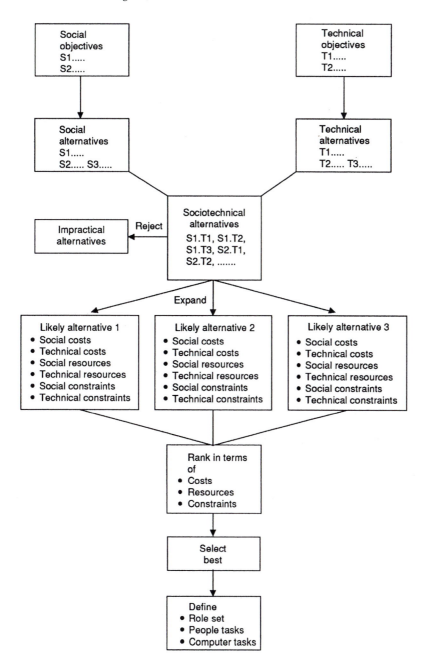

Figure 4.11 Matching social and technical alternatives. (Source: Avison and Fitzgerald, 1988, *Information Systems Development: Methodologies, Techniques and Tools*, Oxford: Blackwell Science, Figure 6.31. Used with permission.)

In practice both Soft Systems Analysis and Multiview are often not carried to completion. The preliminary stages of Checkland's methodology are useful ways of understanding the institutional context of a system but the methodology is insufficiently concrete for the nuts and bolts stage of analysis and design. Similarly, in practice, the final stages of Multiview, where there is little practical guidance on how to reconcile social and technical aspects of systems design, are often not completed.

Reconciling Viewpoints and GIS

'From an IS perspective the most difficult problems have already been solved by the time when requirements are specified so the program design can start.'
Hirscheim and Klein, 1992.

Soft systems methodology as proposed by Checkland and his colleagues is a general purpose problem structuring technique. But it can clearly provide a methodology for helping Geographical Information Systems analysts to probe more deeply the critical question of users' requirements. As Checkland, himself, pointed out with regard to Information Systems in general:

'... as with all problems of information provision, the first stages are nothing at all to do with data, hardware or software. They concern perceptions and politics, the interpretations of their world by the organisations in question; they concern the meanings attributed to the flux of events and ideas through which the organisation lives....

... when [soft systems analysis] produces models which are widely agreed to be relevant in the company situation, then such consensus activity models can be converted into information flow models and the more traditional methods of information system design can be initiated.'
Checkland and Scholes, 1990.

In effect Checkland is suggesting using soft systems analysis as a front-end to more conventional methods of Information Systems development in order to ensure that the systems which are developed are ones which the organisation recognises that it needs and this is why it seems attractive for GIS.

Multiview (Wood-Harper, Antill and Avison, 1985) is an Information Systems Development Methodology which has attempted to take this idea forward by fusing a soft systems approach with more conventional data analysis techniques. It is literally a 'multi-view'ed approach to development, approaching the Information System development process from a number of different perspectives. It should therefore be appropriate for dealing with the multi-disciplinary dimension of GIS implementation which involve different types of uses and often has significant technical implications for both IT and non-IT staff.

Multiview has a number of characteristics that recommend it as being potentially valuable as a corporate GIS development methodology where there are likely to be considerable doubts about the need to introduce such novel technology. Any corporate GIS system is likely to disrupt many long-standing manual map-handling processes and cause some apprehension among 'drawing-board' staff. Multiview's insistence that social concerns be given explicit consideration, and that any proposed system should represent an appropriate balance between social and technical considerations (*not* necessarily an 'optimal' technical solution), would provide a sympathetic procedure for considering such sensitive issues.

The emphasis upon prototyping (see below) and concern to develop appropriate user-interfaces also recommends the methodology for use with GIS projects. Indeed, in a GIS variant of the Multiview methodology, perhaps Stage Four might be extended to consider not only the conventional aspects of user-interface design — menus, help boxes etc. — but also might be concerned with determining appropriate presentations of spatial data — map styles, key designs, legends, zooming and panning controls etc. Multiview is particularly appropriate here as it has been used previously in projects which have been based on software packages, rather than developing bespoke software, and as such it has been suggested as a particularly promising model for GIS Development Methodologies (Hobson, 1991).

CONCLUSION

ETHICS and Multiview are the prime examples of methodologies designed to address the people issue in IS development. We know of no documented examples of their application for a GIS. We have, however, several examples of situations to which they might usefully have been applied and in which some attempt to address the people issues has been made.

One such illustration of the context in which GIS is introduced into a large organisation is provided by the experience of the UK Coal Authority (Offen et al., 1997). GIS was introduced to deal with a vast array of searches (1000 per day) of archives of 100,000 plans and 1,000,000 primary geological records. A fast track automated Property Location System was developed to meet this need but the point is that the project introduced major changes in working practices coupled with a concentration in skills and adjustments in staffing. Eventually users gained experience and confidence with the system and are more willing to accept change in the future.

Such far-reaching changes are typical of GIS because of the business and human resource implications of introducing working with spatial data and spatial processing. Not only are there likely to be, as in the case of the Coal Authority, changes in data structures, data models and concomitant changes in business process, but also changes in working practices and roles of personnel (Aybet, 1996). Such circumstances are likely to test any development methodology since as Scott Adams says:

'People hate change and with good reason. Change makes us stupider, relatively speaking. Change adds new information to the universe; information that we don't know. Our knowledge goes down a tick every time something changes.

On the other hand, change is good for people who are causing the change. They understand the new information that is being added to the universe. They grow smarter in comparison to the rest of us. This is reason enough to sabotage their efforts.'
Adams, 1996.

For a new system to be accepted, there has to be acceptance by the users on the one part and a consideration of users' needs on the other. As any manager will tell you, getting these things right is not easy. ETHICS and Multiview provide two ways in which these problems can be approached. Both methodologies incorporate in the design and implementation stages of IS development:

— values of human systems
— complex relations
— multiple objectives
— organisational culture and politics.

ETHICS in particular addresses these issues head on, and indeed it forms the basis for this particular aspect of Multiview. Mumford (1995) cites many examples of its success but the procedure in both methodologies still leaves key questions unanswered. Jayaratna (1994) sums up the problem succinctly:

'.... the methodology offers many design guidelines useful for the understanding and design of human-centred systems, but.... it does not provide any models for handling the interpersonal and political conflicts that may arise from opening up human feelings and emotions. In effect the success of the transformation, depends very much on the commitment of the participants, and, more importantly, the interpersonal and political skills of the methodology user. While ETHICS is committed to a particular ethical stance it does not offer a means of discussing or resolving many of the ethical dilemmas that could arise in systems development. In effect the creator of ETHICS has not abstracted the knowledge, models, skills and abilities of herself and made these available to potential...users.'

A similar point can be made about Multiview when it comes to the pivotal point of reconciling the multi-views under consideration. No rules or approaches to this problem are provided in the method. It is up to the key player, or players, to have mentally assimilated the results of each stage of the analysis to reach judgements about how conflicting issues are resolved.

For both methodologies therefore, there is no metric or any secure means for balancing or resolving conflict. In the end deciding on the importance of disparate values is a question of individual or collective judgement.

So we see that the real significance of these methodologies is not in laying down rules or procedures which ensure that people matter. Their importance is in formally and significantly putting people in to development agendas. We can, in this light, appreciate that both methodologies can be partially used to good effect in any IS development scenario and there may be no need to follow them through to completion if the people issues are part of the design culture.

SELF-CHECK QUESTIONS

1. What are the three 'drivers' which are identified as motivating the search for new ISDMs?
2. What are the underlying principles of ETHICS?
3. What characterises hard and soft systems?
4. What are a 'rich picture' and a 'root definition'? Attempt to draw a rich picture for a system familiar to you and produce a root definition for the system.
5. How in the Multiview method are technical and social issues resolved?

WHAT YOU HAVE LEARNT IN THIS CHAPTER

* *   *there are alternatives to the traditional waterfall model of system development*
* *   *we can understand the forces which have shaped the development of alternative methodologies as complex drivers which provide both a rationale and design principles*
* *   *the key drivers are that people matter and business is moving faster and technology is affecting both*
* *   *the methodologies can be understood also in terms of the framework of ideas which lie behind their philosophies and methods*
* *   *a common element of modern ISDMs is participation by users*
* *   *this finds its fullest expression in ETHICS whose methods explicitly involve users and their values*
* *   *a key issue is how socio- and techno-centric views are reconciled*
* *   *Multiview, based on Soft Systems Methodology, attempts to do this*
* *   *such socio-technic methodologies address the issue of involving users but give no rules for making judgements*

GOING FURTHER

Texts on comparative methodologies are given at the end of Chapter Five.

Chapter Five: Alternative Information System Development Methodologies: Rapid Development Methodologies

In this chapter we consider:

* *a number of modern methodologies which focus upon rapid development*
* *the ways in which new technologies have allowed the use of prototyping in system development*
* *the importance of prototyping for GIS development*
* *a development methodology, the Organic Life Cycle, based on prototyping and on a model of work called the Spiral Model, which focuses on meeting requirements*
* *a development methodology, Evolutionary Delivery, which uses new technology to focus on the delivery of software products*
* *an approach to system development, Rapid Application Development, in which both methodologies can be used and which focuses on rapid delivery*
* *the problem of selecting a methodology*
* *matching ISDMs to organisations*
* *the comparison of ISDMs on the basis of different criteria*

'The Life Cycle Approach.... should be more properly called the Death Cycle... since it is clear that the hope offered.... that systems developed by this method could subsequently be changed and kept up to date, was spurious.'
Anon, quoted by Grindley, 1993.

'I would like to think it was an educational problem. Unfortunately it is not that simple. Because we have no teachers. We are travelling in unexplored territory, and, since no one has been there before, it's very dangerous to listen to anyone who says he knows the way.'
Anon, quoted by Grindley, 1993.

'As a ranking curmudgeon in the field, I talk to a lot of people out there. Everyone I talk to — without exception — says that this schedule problem is the biggest plague of the field. Not fast-changing technology. Not new management philosophies. Working to impossible schedules is the biggest problem in IS.'
Robert L. Glass, quoted in McConnell, 1996.

RETREAT FROM THE WATERFALL MODEL

The waterfall model was based on a number of assumptions:

— that all its stages could be completed in sequence
— that the costs and benefits of an IS could be calculated in advance
— that users knew what they wanted
— that the work needed was known and could be measured
— that programs once written could be altered
— that the right answer could be produced first time.

With information systems none of these has been shown to be true. Information systems development has proved difficult to manage because it presents three major problems which cannot be dealt with by traditional management methods. These are (Grindley, 1993):

— the difficulty of stating what is required
— the difficulty of measuring pioneering work.

The traditional Life Cycle approach effectively ignores these issues. Solutions to similar problems in mainstream business have taken a number of forms but foremost amongst them are these four:

— contingency planning
— prototyping, running trials on parts of systems
— segmentation, breaking systems into smaller components to reduce complexity
— user responsibility.

None of these solutions are new. What has been new in the last ten years is the translation of these management responses to formal structured methods. What has made these approaches promising is a combination of a climate of opinion which regards the traditional life-cycle as a mistake with the availability of tools to achieve rapid, relatively inexpensive software development. One aspect of the response is the development of socio-technic methodologies but another aspect is the emphasis on ***prototyping*** and the focus on business objectives in information systems development.

PROTOTYPING

A prototype is a working model of a system or part of a system which may emphasise some specific aspects of it.

Prototyping is the mechanism by which ***participation*** is ensured and through which the user's view of the system can be expressed. Prototyping is a response to the problem of not knowing either what the user wants or what he will do with it when he's got it!

Bestebreurtje expresses the problems prototyping addresses succinctly:

'Perhaps the most difficult problem in a project is to define what has to be built. In order to do this the client has to tell what the GIS system should look like. But in practice: *'They do not know!'* because for them it is the first system of its kind they have experienced and they lack the knowledge which comes from experience.'

'Human beings almost never perform a complex task correctly the first time. However, people are extremely good at making a mediocre beginning and then making small refinements and improvements.'
Bestebreurjte, 1997.

Prototyping has long been recognised as an approach to systems development that has been applied to everything from aircraft to zips. It has been mentioned as part of structured systems methodologies for a long time but was never regarded as a major area of activity in systems development, more as a means of trying out new and risky ideas. The trouble with prototyping was that it was time consuming, expensive and led down uncertain pathways. All that has changed in the last five to ten years for information systems development because of changing technology. Software development tools which have become available in the last few years mean that prototyping has become one of the main vehicles for systems development and has begun to revolutionise the whole field of development methods. We shall examine two methodologies put forward in the 1980s which address the three issues of management identified by Grindley (1993) and involve the four solutions to a greater or lesser degree. Each has its own emphasis which can be put down to the practical experience of their authors. But the clearest message for us is one of a convergence of these methodologies and the maturing of an approach to systems development based on rapid delivery.

The implications for GIS are enormous. The special characteristics of GIS, complexity, accessibility, multi-disciplinarity, the newness of the technology, the newness of handling spatial data and the difficulty of the technology all make GIS development fertile ground for a prototyping approach. If we are to understand why prototyping is important for GIS then we should understand something of the technology and something of what prototyping means today. The importance of these topics is easily missed and so we shall make a brief excursion to familiarise ourselves with them before returning to the main question of methodology.

The Technology Behind Prototyping

CASE Tools: CASE means Computer Aided Software Engineering. CASE tools are programs that write code. CASE tools semi-automate the methodologies, modelling and techniques used in system analysis and design. They increase the productivity of design teams and help ensure quality. Typically in a CASE-based system there is a Central Repository, a sort of database management system which is capable of storing, co-ordinating and producing models derived from modelling tools, project management tools, documents, software code, module and object libraries of re-usable code and features for reverse engineering, re-engineering and restructuring of software.

The repository supports the various CASE tools. These CASE tools can be used to handle the main project analysis tasks such as producing entity relationship diagrams, data flow diagrams and screen interfaces. Models can be changed and added to easily without the need to restructure complex diagrams manually, so data processes and entities are not 'etched in stone'. The tools can also check models or errors, consistency and completeness.

The benefits of such tools are in adding discipline to systems development and improving quality, reducing errors and reducing rework. The first major impact on methodology is the facility for prototyping. This is the basis for rapid development and for methodologies such as the *Organic Life Cycle* and *Evolutionary Delivery* which are described below.

A problem with CASE tools and the approaches to systems development they provide are cost of software and a continuing emphasis on the analysis and design stages of projects.

> 'By encouraging development teams to shift their efforts in to the analysis stage they have brought about the dreaded "paralysis by analysis" syndrome which in 1989 one user described thus: "we spent nearly two years doing detailed analysis and went into such great depth that in the meantime the business picture had already changed." '
> Goodwin, 1995.

One response to this dilemma, and an increasingly important aspect of CASE technology, has come from technology itself in the form of Object-Oriented systems and methods. These allow the design and management paralysis to be addressed in a new way.

Object-Orientation (O-O) is a fundamental principle in the design of systems which treats the things that are dealt with as *objects*. An object is something that has a certain behaviour. In information systems attributes and operations are *encapsulated* to create objects. O-O system design is concerned with identifying objects with properties that receive *messages* and undertake certain functions. An object responds to a message, acts on its own and sends messages. It can be made to appear to have intelligence.

Objects have to belong to hierarchical classes and as such can have *inheritance* of properties. Additionally several types of objects can respond to the same message and to respond appropriately. This property is called *polymorphism.*

These properties of objects, as opposed to say entities in conventional data models, give them special characteristics, which are having important impacts on IS design and maintenance. The main benefit of object orientation, which is a direct and specific consequence of an object approach, is *re-usability* of components. In this way software engineering can emulate hardware engineering and achieve higher levels of quality and efficiency and greater effectiveness in use of resources. *With re-usable components it is not necessary to design and build systems by writing all software modules from scratch.*

If we can characterise the approach based on objects as object-oriented then the conventional approach to design can be called structure-oriented. In some senses there should be little conflict between the approaches since they are two routes to achieving the same end. In fact the classic structured methodology (SSADM) now includes O-O in its methodology (CCTA, 1995) with the specific aim of re-use. O-O has a profound impact on the whole design process because it allows a bottom-up approach. The structuring of data into reusable components means that a strongly hierarchical structure can be ignored and emphasis given to component development. In this way the whole resource base, timing constraints and objectives of IS design and implementation are shifted. The problems of control and the weak links in the chain of traditional data model design are altered so that *delivery of usable systems* can take precedence. Using Grindley's (1993) terminology, object orientation allows segmentation of IS.

What O-O thinking and development tools have meant for GIS is that there has been a shift in GIS systems development away from algorithms to customising. New projects are rarely involved in programming or in developing algorithms. Most of the GIS and database functionality required by users already exists and so most business-specific functions can be created by combining existing functionality. Modern system development focuses therefore on customising existing packages. This means creating high level functions which support workflows or major task sets, customising interfaces, linking different software products and fine-tuning performance.

Open Systems: O-O thinking with the principle of re-usability has been a driving force behind the idea and the practice of *open systems*. If objects are the basic building block of software and objects can include not only data but also processes then the effectiveness of objects can be increased tremendously if they can be transferred or accessed between software products, or between operating systems or across networks. The possibility is opened up of having a completely seamless set of information processing modules with an emphasis on 'best of breed' for particular processing algorithms on customisation, on modularised user-oriented processing. Such possibilities are beginning to be realised both in the wider IT world and in GIS and they will have major impacts on the design creation and use of IS (Glover, 1996).

Open systems are defined in many ways and the confusion in definition is not least a product of their politics. The bottom line of open systems for organisations who sell various types of systems, including operating, networking and database systems as well as applications systems and who developed them is that being open could mean being vulnerable. Until recently most software systems were proprietary, that is the source and binary codes which constituted the key software components of any system were encoded and protected by legal devices. With the proliferation of networks, more openness became necessary when it was clear that no single vendor or developer could hope to meet user requirements.

However, to move to open systems the first key issue is discarding proprietary codes and accepting standards which allow different operators with different software to use each other's code, or share data. Open systems ideas are rapidly gaining ground in the late 1990s with the development, publication and acceptance of standards, publication of proprietary formats and setting up of 'open systems groups'.

Another GIS 'Joke'

The nice thing about standards is that there are so many of them...

One possible *de facto* standard is based around Microsoft's Windows environment. Software components within Windows using Microsoft's Component Object Model (COM) and their integration standard (Object Linking and Embedding — OLE) now make vendor-independent plug-and-play systems a reality as long as they conform to these standards (Glover, 1996). With the publication of OLE4GIS specification for spatial data based on OLE/COM, users will have seamless access to data stores using OLE servers within Windows environments which integrate GIS with word processing, spreadsheets, email and many other of the Windows-based products. Customisation tools such as Visual Basic, Delphi, Visual C++ and systems integrators will allow users to create specific business IS environments that can communicate geographically within and between enterprises.

What all this means for development methodologies is that there is a shift from development based on coding to that based on components. Re-usability, not only within projects, but also using software tools provided by vendors and by other enterprises will replace 'build from scratch'. The associated release of resources will then both allow and force a shift in the development to usability and business benefits and away from technical requirements.

In a world where different software products in different languages on different platforms and with different operating systems can be made to work together the second key issue is creating interfaces, or structures in software which can be 'hooked' into by any other software. These interfaces are the key to, and in fact can be seen to define, open systems.

Types of Prototyping

The idea of prototyping is included in many design methodologies. It is part of the techniques of SSADM and is included as part of ETHICS and of Multiview but in all these it has a relatively minor role. Allen (1991), in contrast, provides another view of a much more important role of prototyping in the *whole evolutionary process of design and implementation.* Philosophically, it represents an approach to design but in reality it is derived from the need to answer the shortcomings of the waterfall model. This it tries to do by involving users and by critical testing of complex systems. In this way we can regard prototyping as both a conceptual and a real link between soft and hard systems approaches, between techno-centric and socio-technic approaches. It can be seen as a relatively value free way to achieve participation in design.

In order to understand what this wider concept of prototyping means we have to explore the meanings of the word. Usually it can mean one of the following:

- a tuning tool to enhance system performance
- a means of establishing a user interface
- quick and dirty approaches to system building
- a strategy for avoiding requirements definitions.

None of these describe either of the two approaches to prototyping covered here which have evolved as controlled approaches within structured methods. These we shall refer to as *essential prototyping* and *design prototyping*.

Essential Prototyping

Prototyping is rarely used to deal with the problem of what a system will *do*. However, with complex system development, before we can tackle uncertainty about, say, interfaces or performance, often we need to tackle the uncertainty over what a system is required to do in the first place. Of course, tackling this uncertainty is just what structured systems analysis is designed to do, but this also is what essential prototyping is designed to do.

Consider Figure 5.1 which shows the Organic Life Cycle model. This is described more fully below but is presented here in order to illustrate the point about essential prototyping. The role of such prototyping is to complement strategic modelling and systems analysis. Note in particular that the model is iterative and cyclical, not sequential. The whole process has to start from somewhere, of course. This can be from strategic or systems perspectives. The objective of essential prototyping is to enhance the model by *experiencing* working parts of it, not to build it from scratch. The role of essential prototyping is to feed the rigorous systems definitions that are achieved in the other processes of strategic modelling and systems analysis.

Such an iterative procedure requires control. In any project the traditional deliverables and milestones of the waterfall model cannot be disregarded. It is necessary to have some sort of life-cycle model for the prototyping itself and some understanding of the classes of essential prototyping. The former is given by a prototyping skeleton, Figure 5.2, which is a fairly obvious sequence of creation, evaluation, modification and completion (Allen, 1991).

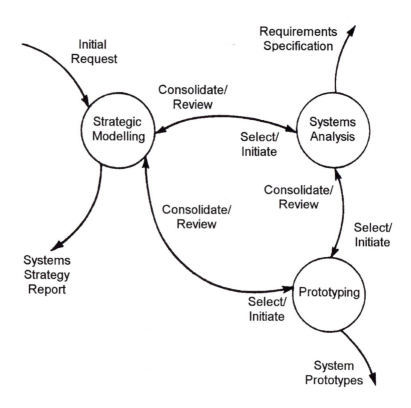

Figure 5.1 Organic life cycle model. (Source: Allen, 1991, *Effective Structured Techniques: From Strategy to Case*, Hemel Hempstead: Prentice-Hall International, Figure 1.3.)

Allen (1991) gives us three sub-classes of essential prototyping:

- *Primitive prototyping*: This is normally a response to strategic modelling as a means of verifying portions of a data model produced from data-oriented systems investigations.

- *Event prototyping*: This tests responses to events in the system.

- *Decision support prototyping*: This involves subjecting a database structure to ad hoc enquiries or tasks required by different levels of management in order to refine the data structure.

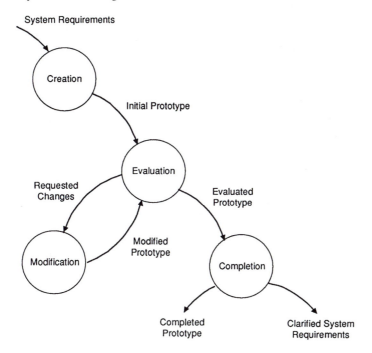

Figure 5.2 Prototyping skeleton. (Source: Allen, 1991, *Effective Structured Techniques: From Strategy to Case*, Hemel Hempstead: Prentice-Hall International, Figure 10.3.)

Benefits and dangers of essential prototyping

The benefit of prototyping is reduction in uncertainty. In situations in which there is little uncertainty there will be few benefits, but in complex situations characterised by uncertainty the benefits of prototyping are:

- **clarification of requirements** because it is real and tangible and because it lends itself to iteration, which is a key element in channelling the learning process. The user is often heard to say 'I don't know what I want until I see it' and prototyping is the vehicle that accommodates this. The fact that users can change their minds is a recognised and indeed encouraged part of the process (Giddings, 1984). As a result the users achieve a better understanding of what they want and of what they are getting.

- **improving productivity** by eliminating the need for what were the later stages of the product life-cycle, particularly testing. In early stages it reduces the learning necessary on the part of analysts, in the later stages it reduces the learning necessary on the part of users. This can result in the earlier delivery of the eventual product. However, it is a common misconception that prototyping reduces costs. Whether it does or not depends on the trade-off between early user input and iteration and late user input and restructuring.

- **psychological benefits** of delivering a system which works, which users are familiar with and in which they have some stake or ownership. Prototyping encourages involvement.

There are dangers in prototyping which are well recognised:

— **the prototype becomes the finished system** either because of inflated user expectations which provides pressure for early delivery or through lack of time through excessive iteration.
— **insufficient preparatory analysis** resulting in poor initial understanding.
— too much attention to cosmetic details especially user interfaces.
— **poor documentation** as a consequence of iteration.

The realisation of these dangers arises not from prototyping but from poor management because of an inadequate understanding of prototyping. A project plan which relies on prototyping should emphasise these points:

— it **is** part of a well-planned strategic approach
— it is **not** quick and dirty
— it complements a wider structured technique
— it is a user-friendly way of enhancing the accuracy of specifications
— there are different types of prototyping
— there are different sub-classes of essential prototyping.

Proper management should reduce or eliminate the dangers by making the place of prototyping explicit in the project. This should specify what type of prototyping is being planned. In addition, resources should be managed to control iterations and emergence of unnecessary types of prototyping. Finally, there should be cultural input, explaining prototyping as part of the project philosophy. Management should avoid the extremes both of conservatism and of radicalism. How the manager pitches the approach depends on externalities such as user temperament, deadline pressures, planning stringency and risk.

Design Prototyping

Within the traditional mind-set of systems analysis, prototyping provides an opportunity to explore alternatives *after* requirements have been rigorously defined. It is not regarded as a tool to establish requirements in the first place. In fact there is some conflict between prototypes and systems analysts based on mutual stereotypes; one of 'toys for the boys' — an uncontrolled activity with no organisation; the other of time-consuming, unproductive, esoteric and mysterious paperwork. Both stereotypes are important in feeding misconceptions, and thus management anxiety. It is necessary to remove some of these in order to see how prototyping relates to structures techniques.

The established idea of prototyping is to spend a limited amount of time and money on producing something in the small before producing something in the large. This reduces uncertainty and minimises overall costs. In most branches of engineering this uncertainty relates to how a system will work. There are such questions as 'will the bridge stand up?' or 'will the plane fly with this payload?' Design prototyping addresses such issues.

Two useful classes of design prototyping can be distinguished:

- user-interface prototyping
- technical prototyping.

User-interface prototyping focuses on issues of screen layout, dialogue style and ergonomics. In fact, this prototyping relates closely to soft systems analysis since issues of usability are allowed to affect requirements specification.

Technical prototyping or simulation is used in detailed systems to determine the adequacy of a solution from a performance viewpoint. It is useful when we are interested in the environment in which a system operates and with which it must interact. This can mean hardware platforms or situations where response time is important. Again there can be considerable linkage with the definition of soft system requirements.

Prototyping and GIS

Prototyping is as we said an established part of engineering practice to which software developers have only recently woken up (though see Dearnley and Mayhew, 1983). That awakening has included GIS probably because the introduction of GIS has profound impacts on business and usually means some sort of *business process reengineering* has to take place. This involves not only the analyst understanding the capabilities of systems in practice but the users going through some sort of education process. Prototyping is essential for this.

> 'Some sort of prototyping, using the productive system or a simpler one, is the only way to really educate prospective users of GIS. Only by being shown what is possible can the users decide what is desirable, which itself must be tested for feasibility and understanding and shown once more.'
> Hobson, 1991.

In a thorough and critical study of the introduction of GIS in the US Army topographical support service Peuquet and Bacastow (1991) put forward prototyping as the key to success where the traditional waterfall model had failed. They recognise the crucial importance of organisational issues for GIS implementation. Their analysis:

> '...strongly suggests three lessons universal to the introduction of (GIS) in large organisations. These lessons are:
> (1) The classical life-cycle development approach does not work well when the project involves the initial introduction of a given technology within any organisation.

(2) There must be an organisational recognition of the need for, and a strong commitment to significant changes in the organisational structure. The determination of this structure and how the GIS is to function within the organisation must be included as an integral part within the development process.
(3) The involvement of the entire organisation is essential for a successful first development effort. No development of GIS can be successful without the information and support that management at all levels, as well as the prospective users, can contribute.'
Peuquet and Bacastow, 1991.

And their solution is prototyping:

'Specifically, what the development of GIS needs is a development strategy that allows the organisation and the people in it to adapt and evolve in parallel with the other components of the system, as well as a composition of the development team that is designed to maximise the effectiveness of that strategy. It is suggested that the best development strategy is iterative prototyping.'
Peuquet and Bacastow, 1991.

Others have argued elsewhere that prototyping is essential 'if GIS is to move from being technology-led to user-led, and if GIS is to reach its full potential in various applications settings' (Dunn and Harrison, 1991).

So it seems that GIS is beginning to get away from the constraints of the traditional approach to development. Recognising the effectiveness of prototyping might however be making a virtue out of necessity. Prototyping is not something done in isolation as a bolt-on to traditional methods or as a polite way of saying 'muddling through'. Prototyping is part of the methodological swing away from the traditional approach to one that is based on software tools and approaches such as O-O technology. The point is that the technology and the methods go hand in hand.

'The successful large-scale adoption of object technology for ISD does not depend simply on the availability of object-oriented languages in a technically well appointed environment. It also depends on the adoption of the object-oriented philosophy or mindset by both the technicians and managers and on the utilisation of a formal methodology that includes project management guidelines.'
Duc and Henderson-Sellers, 1995.

Methodologies have developed which are built on the idea of prototyping and the real lesson for GIS development is to learn what these are about. One such is the Organic Life Cycle put forward by Allen (1991).

ORGANIC LIFE CYCLE (OLC)

Duc and Henderson-Sellers (1995) emphasise the importance of methodology to control the use of these powerful new technologies. There presently is a period of adjustment of techniques and methods to the fluid, complex reality of systems development and little time to see the emergence of a mature methodology. One methodology, articulated as a response to the impact of CASE tools which has relevance for the later developments based on object-based technology and open systems, is the Organic Life Cycle (OLC) based on the idea of prototyping. Where OLC begins is with the problems of sequential structure in established formal methods which are based on the waterfall model.

The Organic Life Cycle (Allen, 1991) is a direct response to the perceived shortcomings of structured techniques in general. The problems Allen identifies fall into two groups: those concerning the failure of structured techniques, and those concerning their application. His response is the Organic Life Cycle model represented in Figure 5.1.

The origins of the OLC lie in *spiral models* of the life-cycle such as that provided by Boehm (Figure 5.3). This is a set of repetitive steps whose iterations are meant to converge to a prototype.

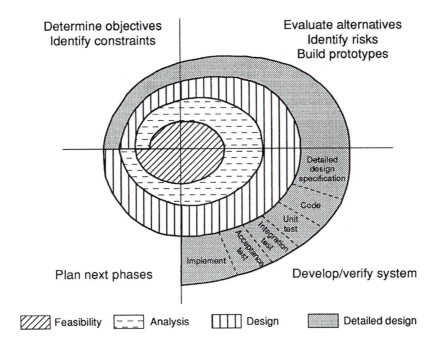

Figure 5.3 A spiral model of software development. (Source: Allen, 1991, *Effective Structured Techniques: From Strategy to Case*, Hemel Hempstead: Prentice-Hall International, Figure 1.2.)

There are four steps in each iteration corresponding to the four quadrants of the diagram. The key element is the idea of iterative activities that are applied with the aim of reducing uncertainty. In the OLC, systems strategy, analysis and prototyping are seen as continuously iterative or cycling, inter-connected processes. Important in the structure and operation of the model is the idea that interactions of these processes are at the project level, i.e. they are carried out as individual projects, completed and results delivered independent of other activities and according to a plan. Also important is the idea of growth, i.e. that interactions, although acting independently and even in parallel, are in some way goal-directed and reinforcing.

The fundamental idea of the Organic Life Cycle is that of growth. However, it is distinct from the waterfall approach and from other approaches in allowing, indeed being designed for, development in requirements and specifications.

Note also that the diagram (Figure 5.3) is incomplete. Projects do not start with a 'clean slate' every time. They need to be selected and initiated in the context of what has already been learnt. This is of course the subject of constant review as new experience and knowledge are built in by iterations of projects. Operation and control of the whole system requires some strategic model for an organisation as a whole. Other activities that relate to organisations, such as implementation, are not shown in order to preserve clarity.

The OLC approach is fundamentally different from other ISDMs in that the underlying models are dynamic and continuous and not sequential. Allen also regards it as a premise of the model that projects are undertaken quickly — they are short term and reviewed, maintained and completed on a regular basis.

EVOLUTIONARY DELIVERY (ED)

Another methodology which has relevance for GIS is Evolutionary Delivery (Gilb, 1988).

In Evolutionary Delivery we see a methodological response to the problems of the *business is faster* driver. The approach doesn't address the social issues in a systematic way as do ETHICS and Multiview, although, it has a strong emphasis on the management of people and on a democratic approach to project implementation. It is essentially about management of project design and its emphasis is on *delivery*.

If we had to summarise its philosophy, we would describe it as being based on the idea that a good project is one which delivered useful products *on time and within budget*.

Gilb's (1988) analysis of information system failure produces a set of principles for software engineering management.

- The invisible target principle — all critical systems attributes must be specified clearly.
- The all-the-holes-in-the-boat principle — your design solutions must satisfy all critical design attributes simultaneously.

- The clear-the-fog-from-the-target principle — all critical attributes must be specified in measurable, testable terms and the worst acceptable level identified.
- The learn-before-your-budget-is-used-up principle — never attempt to deliver large and complex systems all at once, try to deliver them in small increments so that you can discover problems and correct them easily.
- The keep-pinching-yourself-to-see-if-you-are-dreaming principle — don't believe blindly in any one method, use your methods and common sense to measure reality against your need.
- The fail-safe-minimisation principle — if you don't know what you are doing, don't do it on a large scale.

These principles in whimsical style encapsulate the experience of the author and the essence of the philosophy. They recognise that most IS projects are large and complex and that *delivering* systems that are successful in business terms is the only criterion that matters. So the emphasis is on:

- identifying exactly what it is you want to deliver
- identifying the necessary conditions for delivery
- make sure you know what delivery means, i.e. what solutions are
- keep your eyes on the target of delivering something whatever the problems.

Much of the explanation of evolutionary delivery is taken up with an analysis of the issues of:

- identifying real problems
- determining what are solutions
- evaluating solutions.

In comparison with SSADM, the emphasis is on requirements. But it is not on the *full* specification of them rather on their *proper* specification. The initial problem with projects is regarded as the lack of management clarity in stating goals.

> 'There is usually no clear distinction made between the results or the goals we must aim for — the *problem*, and the various possible courses of action for getting those results — the *solutions*. We are constantly making the mistake of specifying the means of doing something, rather than the results we want. This can only limit our ability to find better solutions to real problems.'
> Gilb, 1988.

In project documentation and approach this adds up to the:

---
principle for the separation of ends and means — avoid mentioning solutions in your goals statement

---

and provides the rationale for an emphasis on *metrics*, i.e. ways of measuring outputs. Metrics are regarded as essential for controlling any project and in Gilb's approach great emphasis is placed on them. Gilb is a firm believer in Kelvin's principle that:

— when you can measure what you are speaking about and express it in numbers, you know something about it, but when you cannot measure it, when you cannot express it in numbers, your knowledge is of a meagre and unsatisfactory kind.

The idea of metrics might seem techno-centric but it is not so necessarily. This principle has echoes in Mumford's ETHICS in the attempt to define and measure even such a fuzzy idea as job satisfaction. Indeed many elements of both this approach and of Multiview are attempts to improve methodology by introducing the rigour of measurement even in a socio-technic environment.

Use of metrics brings advantages at all stages of analysis and design which relate strongly to issues of motivation, participation and organisation which are incorporated in the socio-technic approaches. Metrics provide more certainty, identicality, clarification and testability for users or clients as well as for designers. For the design team they provide estimatability, resource control, design by objectives and means control.

The next stage of analysis is to clarify what are *solutions* for a project's requirements. A solution is a 'set of ideas the implementation of which impacts on at least one part of the problem positively', (n.b. this implies we know exactly what the problem is and we can identify what is positive). What this means is that a solution contributes to at least one function, or attribute, requirement. A solution with no positive effects is not a solution. Some solutions have negative effects. Gilb uses the famous swing example presented in Chapter Three to make the point.

His point is that most people think the point of this figure is to do with 'communication' or 'documentation'. True, but this is not the real problem. The figure has to do with the poor specification. In all drawings the basic specification is understood — the differences are in the quality of the specification and the corresponding resources used.

A key issue in specifying solutions is understanding the difference between *solutions* and *goals*. Goals can be functional, such as 'I want to be an academic', or attribute, such as 'I want to be rich'. Solutions are the means to achieve this. Solutions must be kept entirely separate from goals. They are easily confused. If in a project it has been decided that you must use a particular solution, then that solution is a *requirement* and it should be included as such from the beginning as a functional requirement. If GIS is to be used to solve a business problem (without debate) then it is not a solution.

This point is expressed as the unholy solution principle:

solutions are never holy, they can and should be changed in the light of new requirements.

The next problem is to evaluate solutions. This is usually complex and uncertain and it is in this area that the value of metrics is felt. Solution evaluation is made in relation to attributes, especially those which are identified as critical attributes. The problems come when there are multiple goals of which some inevitably are in conflict. On top of that it is normally impossible to quantify the contribution of any particular solution.

The first principle is to deal with uncertainty. The uncertain certainty principle:

uncertainty must be stated in no uncertain terms

allows designers to focus on those areas where evaluation is needed and where comparisons of solutions must be made.

This emphasis on accurate specification (rather than full specification of traditional methodologies) on metrics and on outcomes is the key to early effective delivery and to the functioning of an evolutionary approach as opposed to a rigid sequential one. Although so much emphasis is given to the discipline of these aspects of systems analysis, Gilb's focus is very much on adaptive management for design.

The main concepts behind the approach are:

- multiple-objective driven
- early, frequent iteration
- complete analysis, design build at each step
- user orientation
- systems approach not merely algorithm orientation
- open-ended basic systems architecture
- result orientation, not software development process orientation.

> 'Phased planning asks a dangerous question: "how much can we accomplish without some critical constraint (budget, deadline, storage space)?" Evo-planning asks a very different question: "how little development resource can we expand and still accomplish something useful in the direction of our ultimate objectives?" '
> Gilb, 1988.

Such an approach requires analysis, design and build at each such step. The benefit this brings is the ability after each step to adjust the solutions or objectives to meet what is required.

This approach can be seen to be a variation of prototyping which forms the main method in the Organic Life Cycle. Like prototyping it helps take care of some of the socio-technic issues of the focusing on usability, participation in design, commitments to success. It does this without the formal socio-technic methods because an evolutionary approach necessarily involves repeated feedback, communication and evaluation. Early delivery means:

- users are happy with your activity and are therefore supportive

- the design team are confronted with the realities of users at an early stage
- if nothing useful can be done this is exposed early
- you can get experience of estimating at an early stage
- sudden budget cuts or other external factors don't deprive you of achieving something.

Delivering a system on time and within budget is only one angle, albeit an important one, of the modern problem of rapid development. Development methodologies such as OLC and ED address specific issues in order to deal with specific risks to systems development. There have been many parallel developments in method also addressing the problems of rapid development and there appears to be a convergence of these methods or approaches into a coherent philosophy that has come to been known as Rapid Applications Development or RAD.

RAPID APPLICATIONS DEVELOPMENT

'The aim is to get something approximately right quickly rather than exactly right slowly.'

Rapid Applications Development (RAD) is a term used by James Martin (1991) to describe a set of procedures and approaches which, as the name describes, aim to produce systems designs quickly.
The four key elements of RAD are:

- *Joint Application Development (JAD)* — that is the co-operation of users and developers to analyse and design a system. JAD sessions attempt to converge designers' and users' views of existing and proposed systems through iterative adjustments.

- *Prototyping* is an essential element of this joint process allowing users to see early attempts by designers and provide concrete, positive responses.

- *CASE tools* allow rapid development and adjustment of prototypes, keeping joint sessions alive and focused and helping disparate groups in complex projects focus on key issues rather than getting bogged down in time-consuming procedures.

- *SWAT teams* are the small highly skilled groups who operate the CASE tools and provide the design capability.

RAD focuses on things like the 80:20 rule — that 80% of users' functional requirements can be met with 20% of functionality in a system. RAD identifies and quickly delivers the critical 20%. This is then used and new lessons learnt and specifications changed which can be accommodated quickly and cheaply by the SWAT team. In other words the project can develop organically and in an evolutionary manner.

RAD is not so much a methodology but more an approach to systems development. It necessarily involves some methodology even by default but its focus is on rapid delivery and what RAD is concerned with are the techniques and tools which will allow that.

McConnell (1996) gives us four pillars of RAD (Figure 5.4).

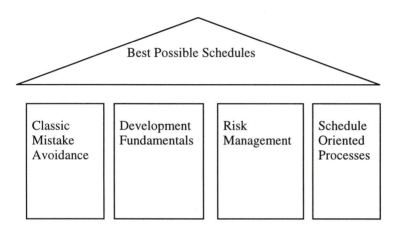

Figure 5.4 The four pillars of Rapid Applications Development. (Source: McConnell, 1996, *Rapid Development*, Redmond: Microsoft Press, Figure 2.1. Used with permission.)

The best possible schedule rests on avoiding classic mistakes, development fundamentals and risk management as well as on sound scheduling. The point is that fast scheduling alone will not produce successful rapid development.

In order to understand what these pillars mean in practice we have to look at the four dimensions of a project which can be managed for speed. These are:

— People
— Product
— Process
— Technology.

People have more impact on software development productivity than any other factor. Differences of 10 to 1 (DeMarco and Lister, 1987) between individuals and of 5 to 1 between teams (Boehm, 1981) are often recorded. Peopleware is the key factor in success and the keys to its success are *organisation of teams* and *motivation*.

Product can be managed for rapid delivery by cutting product size (the 80: 20% rule works!) or by changing its characteristics.

Process can be managed for speed by avoiding reworking, targeting resources and focusing on user needs. But it rests also on selecting the appropriate life-cycle model: the commonly agreed plan about what the plan of action is and how all the bits fit together. In other words the development methodology. No point in selecting the waterfall method for rapid development. You are forced in to something like the Organic Life Cycle or Evolutionary Delivery (more on this later since the selection of the proper method is a wider issue simply than that of RAD).

Technology can be used through appropriate selection of tools. The move to componentware has an important influence here. The problem in each project is, which of the four dimensions is critical in the development process? The project should be analysed 'to determine which dimension is most limited and which you can leverage to maximum advantage. Then stretch each to the utmost. That, in a nutshell, is the key to successful rapid development' (McConnell, 1996).

Doing a lot of things right, however, doesn't make rapid development. You also have to avoid doing a lot of things wrong. There are classic mistakes to which all projects are prone which must be avoided for rapid development (Figure 5.5).

A glance at McConnell's lists in Figure 5.5 shows how many of these mistakes are down to people and their agendas. We shall pick up on this theme in Chapter Six on peopleware. What we emphasise here however, is that even the technology and process related mistakes are the consequences of socio-technical mismatches and false expectations, not of method nor of machine.

Rapid development exposes the weaknesses in any project and it is essential, therefore, to get the basics of project development right. These are the management fundamentals of estimation and scheduling, planning, monitoring and control. Also the technical fundamentals of design and implementation have to be sound. In rapid development, quality assurance becomes critical since the time taken to fix mistakes can be greater than the time required to code in the first place. Quality assurance is based on staff performance and on systematic checking and rectification. The point about defects is that they are best identified and rectified early. A requirements defect, i.e. not getting the customer's needs right, can cost from 5 to 200 times more to fix than it would have done at the time requirements were being specified. In other words, the fundamentals can be ignored only at the developer's peril.

Problems will always occur in projects for one reason or another. But, as much of what has gone earlier emphasises, there is a lot of experience of what can go wrong. The problem for the rapid developer is to identify and manage the risks affecting the project. Risk management consists of a set of activities which seek first to identify the nature of risks which might affect a project, then to prioritise those risks in terms of the timing and magnitude of their likely impact and then to select and respond to the risks in an orderly fashion, all the time monitoring the changes in the priorities. There are techniques for risk analysis which are beyond the scope of this text but the important things to note are the roles of non-technical activities in risk management. Risks are not absolute. They are recognised by individuals and groups and are appraised. Risk analysis is very much a value-oriented process:

> 'It is not only the magnitude of the risk that we need to be able to appraise in entrepreneurial decisions. It is above all the character of the risk. Is it, for instance, the kind of risk we can afford to take, or the kind of risk we cannot afford to take? Or is it that rare, but singularly important risk, the risk we cannot afford not to take — sometimes regardless of the odds?'
> Peter Drucker, quoted in McConnell, 1996.

People Related Mistakes	Process Related Mistakes	Product Related Mistakes	Technology Related Mistakes
1. undermined motivation	14. overly optimistic schedules	28. requirements gold-plating	33. silver-bullet syndrome
2. weak personnel	15. insufficient risk management	29. feature creep	34. overestimated savings from new tools or methods
3. uncontrolled problem employees	16. contractor failure	30. developer gold-plating	35. switching tools in the middle of a project
4. heroics	17. insufficient planning	31. push-me, pull-me negotiation	36. lack of automated source code control
5. adding people to a late project	18. abandonment of planning under pressure	32 research oriented development	
6. noisy crowded offices	19. wasted time during the fuzzy front end		
7. friction between developers and customers	20. short-changed upstream activities		
8. unrealistic expectations	21. inadequate design		
9. lack of effective project sponsorship	22. short-changed quality assurance		
10. lack of stakeholder buy-in	23. insufficient management controls		
11. lack of user input	24. premature or overly frequent convergence		
12. politics placed over substance	25. omitting necessary tasks from estimates		
13. wishful thinking	26. planning to catch up later		
	27. code-like-hell programming		

Figure 5.5 Classic mistakes in systems development.

Finally, rapid development rests on appropriate scheduling. This is the nub of the rapid development approach. The developer under pressure from clients and managers is apt to make over-optimistic estimates. Human nature often tries to keep everybody happy and rapid delivery makes people happy. But unrealistic scheduling is today the major cause of software development problems. However, there is a paradox here. The RAD environment means prototyping, evolution of ideas and an organic development of product but all these things are intrinsically difficult to schedule. Scheduling techniques are available and experienced developers can provide good estimates of time and resources based on past projects but in a RAD environment these have to change. The problem with scheduling is thrown into the area of risk management and the problem for the developer is how to beat schedule pressure.

Three factors tend to contribute to poor scheduling:

- wishful thinking
- little awareness of estimation and of over-optimistic scheduling
- poor negotiating skills.

The key to negotiation is separating people from problems and trying to identify win-win situations or conditions. The developer needs to focus on the real objectives of the project, on results not solutions and on interests of the main parties not their positions. Negotiation and re-scheduling also demand steely nerves:

> 'make no mistake — rapid application development ...is not for the lily-livered. It takes determination, hard work and a willingness to consign everything you knew about application design to the scrap heap.'
> *Computing*, November 1995.

In other words the crucial area of RAD focuses again on peopleware: on skills of management and interpersonal skills of people. It has more to do with positions in and between organisations and how these are handled in a business context. When these things are in place and work, then the benefits of technology and improved methodology can be realised. But not until then.

It is worth noting that like ETHICS and Multiview, OLC and RAD provide no advice or procedures for these crucial, people-dependent aspects of methodology. Gilb (1988) stands alone in building into his strategy some aspects of this knotty issue. He covers two specific key topics for RAD: deadline pressure and motivating colleagues and shows how the methods he advocates, using metrics and focusing on early delivery, contain the hooks by which people can be motivated, encouraged to work together and reach consensus.

The Current Challenge from Technology

Methodologies must be understood in the context of prevailing technology. The traditional methods were devised in the period when coding was done manually, even at machine level, when processing times were *one or two orders of magnitude* slower than today, when it needed large rooms with very expensive air conditioning systems to keep the machines cool and when very few people in an organisation had access to terminals. All that has changed.

Today, high-level languages, database technology and software tools have released systems developers from the constraints of traditional structured methods and allowed responses to the issues of user involvement, user-led design and business process reengineering. The methodological changes which have paralleled these changes are recorded above in the socio-technic methods and the technologically based methods which now dominate the IS world.

But the changes in technology continue apace and as we enter the last few years of the century we are set to witness a revolution in IT whose implications for GIS and for GIS development methodologies are impossible to articulate with any confidence. Remember that the first prediction of world-wide sales for the IBM mainframe computers of the 1960s was a total of five! And who five years ago would have predicted the impact of the Internet?

What we can do is merely to outline some of the changes taking place and to speculate freely on their likely impacts. We can tentatively try to work out how development methodologies might need to evolve and where the emphasis in systems development might be in five years' time. But don't believe it implicitly.

The vision of the near future for GIS is made up of open systems based on interfaces between different data sets, data warehouses, componentware, applications languages, semantic translators and of course the Internet, though a vastly expanded one (Glover, 1996). This vision is being driven by the possibilities of business benefits that can be derived from access to data. It is far from being a dream. It is being formed in people's minds through such concepts as the National Geospatial Database being championed by the Ordnance Survey and similar concepts of a European Geographic Information Infrastructure (or European Policy Framework for Geographic Information) which would be:

> 'a stable, European-wide set of agreed rules, standards and procedures for creating, exchanging and using GI.'

An Open GIS Consortium has been set up by major vendors to promote such ideas and to realise the technologies that will enable it. The first goal is to develop open systems based on open, published interfaces between what were proprietary data structures. This means that a single interface can take data from databases created by different software systems and use them as though they were all in the same format. *Interoperability* of systems will be achieved by a standard object handler which can work from standard operating systems such as Windows. These handlers will allow the development of componentware, that is collections of software routines which access objects in common. These components need not all be produced by the same vendor nor need each other to function. They will be 'best of product' applications which the user can mix and match at will. Their power will be in using data seamlessly between routines and the embedding of one function within another with full functionality.

The functioning of such systems will depend for efficiency of operation on data warehouses which reduce duplication and can manage long transactions. This means that many users or applications can access databases that need only have single copies of data objects. What this means from the business perspective is that GIS can be seen and used as part of what the vendors would call a 'total solution'. GIS can become part of a general information strategy and specialism will evaporate as all users gain access to all software.

A key element of such a vision is the development of interfaces. This is not simply a question of producing neat point-and-click functionality with pop-up menus. It is more a question of ergonomic design of the IS governed by workflows, user comfort and effective outputs, which in the crude business sense translates into cost-effectiveness. The linking of different software components and the creation of effective interfaces is achieved using some sort of customising toolkit or application language such as Visual Basic or C++ or Delphi. Much of the work of the future GIS or GIS in IS will be taken up with the use of such applications languages for customising interfaces rather than the coding of either data processing routines or the creation of system specific architecture.

At the same time the ability to access different databases and different data models will shift the burden of database design and data capture. Access to data will become more significant and data quality will be paramount. At the same time data provision by third parties will become more significant as data is shared between systems.

We can appreciate that the emphasis of development methodologies will shift in parallel to these technology-driven changes. What these shifts will be is difficult to say but we can speculate that there will be greater emphasis on user skills. If users are to make effective use of componentware then their skills may be the limiting factor in successful applications. Then training rather than being an add-on at the end of a project in order to get staff acquainted with new systems will become the key issue in project initiation and design. If data is more accessible to all users in and between organisations then in the socio-elements of systems design communication between groups and organisations will become more important. The design stages dealing with people will need to address this issue specifically and to an increasing degree. Roles will change because of it, as organisations become more transparent to the people who work in them and there will accordingly be a need to consider to a greater degree the linkages in organisations. All these changes point towards corporate approaches to information systems and they will intensify the demands for ISDMs which deal with the tensions which go with a corporate approach, which RAD and prototyping seem to be addressing.

Whatever happens we can be sure that change is on the way and systems developers will need to adapt their tools and opinions if they are to succeed in building systems to meet business needs. There will always remain the problem, however, of what methodology to choose and follow? Whether or not to take a RAD approach or a more traditional one?

MATCHING ISDMs TO PROJECTS

> 'Many have dreamed up republics and principalities which have never in truth been known to exist: the gulf between how one should live and how one does live is so wide that a man who neglects what is actually done for what should be done learns the way to self-destruction rather than self-preservation.'
> Niccolo Machiavelli, 1519.

Our comparative analysis of methodologies began with the idea of failure of the waterfall model. It emphasised the philosophies of alternative methodologies and showed how particular drivers lay behind them. It is clear, however, that the methodologies are not completely distinct. They are not the simple product of a single influence. None is wholly techno-centric nor wholly based on soft systems. Neither do they address single issues. Each has a role and a value. They are not, to paraphrase Gilb, either completely holy or completely unclean.

The real problem of alternative methodologies for GIS both now and in the future, is not 'what is the best methodology', nor perhaps 'what is the best methodology in this *situation* for this *purpose*?' but maybe nearer to 'which methodology is adequate to produce business benefits?' This is the question we want to address in general terms in this final chapter. Approaching this problem requires an understanding of two issues:

- the characteristics of organisational situations of GIS
- the characteristics of the methodologies.

Organisational situations: Organisations can be looked at in many different ways but in terms of the issues confronting IS analysts and designers the two most important issues probably are:

- how far the organisation has evolved an IT culture
- how the organisation's different groups work together.

The evolution of an organisation in relation to IT can be described as four stages of IT impact (Figure 5.6). The idea is that if technology is to succeed in an organisation it must be appropriate for the particular type of organisation. If IT potentially has a high business impact and large effect on business processes then to succeed it needs an organisation in which business processes are highly structured, compartmentalised and whose divisions undertake only specific tasks. Other types of businesses are unlikely to benefit from the sort of IT (such as GIS) whose rationale is based on business processes and whose operation necessarily cuts across boundaries.

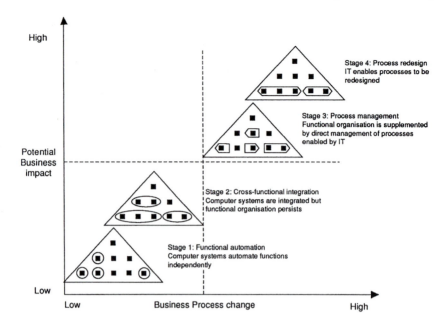

Figure 5.6 Stages of IT impact. (Source: Bestebreurtje, 1997, *GIS Project Management*, unpublished MSc, Manchester Metropolitan University.)

Figure 5.6 shows in one sense an 'evolution' from divisional to corporate operations and thinking. GIS is, as has been argued elsewhere, a technology which almost assumes a corporate approach. So, if a GIS project is undertaken in an organisational structure which is not aligned in such a way — Stages 1 and 2 of Figure 5.6 — then no matter how well the project is managed, it is likely to suffer problems and will fail even if completed according to plan. The benefits of GIS will be realised only in businesses which have business processes which are cross-organisational (stages 3 and 4). Then the power of geo-referenced data and visualisation of complex data sets can be effective.

Another idea about organisational structure and function has been expressed by Howard (1993) who talks in terms of *'client complexity'*. This refers to the degree of agreement among users with regard to their requirements. *Unitary* groups of users share common values, beliefs and interests. They are likely to agree ends and means and then act in accordance with their agreed objectives. *Pluralist* groups will share some common ground, perhaps agreeing in principle that some sort of system is required, but there may be considerable differences about the details of the system which is needed and how the project should be progressed. *Coercive* groups of users will not share common interests, will find compromise difficult and may well disagree fundamentally about whether a new system is required at all.

Howard relates these types of 'client' to *'system complexity'* as a basis for selecting a methodology. 'Simple' systems involve little technological challenge, whereas 'Complex' systems demand technologies which are new to the organisation. Figure 5.7 presents a selection grid which reflects some of the methodologies discussed in this book. Flynn (1992) similarly describes how variables including uncertainty about the user characteristics, about process requirements, and developers can be used as axes to help characterise the nature of a project and thus to determine an appropriate ISDM.

	Simple	**Complex**
Unitary	End user Development Application Generators **'Desk Top GIS'?**	Waterfall model SSADM ED OLC
Pluralist	Prototyping ETHICS ED OLC	SSADM ED OLC **'Corporate GIS'?**
Coercive	Soft Systems Analysis ETHICS ED OLC	Soft Systems Analysis Multiview ED OLC **'Corporate GIS'?**

Figure 5.7 Methodology selection grid.

The positioning of the various methodologies is a simple reflection of their underlying objectives and their scope, as determined by the procedures and tools used in each method. The grid provides an understanding also of how GIS systems might be classified. The GIS research application might well fall into the Unitary/Simple category. In such a situation then the appropriate methodology would be one in which there is close relation between developer and user with the focus on turning out applications directly. Some departmental GIS systems, such as in a local authority Estates Department might fall in the Pluralist/Complex category. In such a situation there could well be different groups who could agree on the need for a GIS-based land information system but who had different needs and different positions within the authority. Here the appropriate methodology might be OLC. Ambitious corporate GIS proposals probably would be located in the complex column but depending on the make up of the organisation would be in the pluralist or coercive categories.

Characteristics of Methodologies: One of the messages of this book is the lack of conscious methodology used in GIS implementations. Our experience of the work of business, local authorities and utilities shows a regrettable lack of awareness of methodology and few cases even where it is recorded. GIS has developed from a niche role in many organisations with small teams (even of one) working with limited resources trying to demonstrate the benefits and power of the technology. As GIS emerges in to mainstream IS and in to mainstream business operations then this situation cannot persist and the methodologies used to implement new systems have to change. Often the methodology used in existing GIS systems is what some call 'muddling through', others call it *'Code-and-fix'* (McConnell, 1996). This is a model of working that is widely used but seldom useful. If you haven't consciously adopted a method then probably you are using code-and-fix. Combined with a tight schedule this gives rise to the 'code-like-hell' approach which stands in contrast to all structured methods. Code-and-fix has the advantage of little overhead and no requirement for expertise. It can be useful for 'proof on concept' work or for small demos. But for real projects it is dangerous. However, we will include it in our comparison because it is probably the most widely used approach to GIS development.

A number of authors provide comparisons of methodologies. Most often (see Avison and Fitzgerald, 1988 and Tudor and Tudor, 1995) this is a comparison of the scope of methodologies in relation to the Business Life Cycle. Figure 5.8 is such a comparison based on the cited works but in addition it includes OLC and ED which they do not consider. The scope of these authors' Life Cycle is extended also to include issues covered in this text.

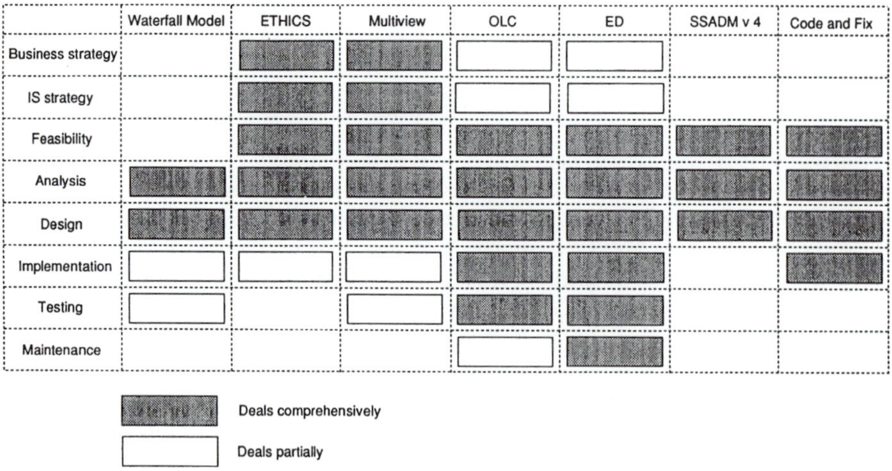

	Waterfall Model	ETHICS	Multiview	OLC	ED	SSADM v 4	Code and Fix
Business strategy		▓	▓	▢	▢		
IS strategy		▓	▓	▢	▢		
Feasibility		▓	▓	▓	▓	▓	▓
Analysis	▓	▓	▓	▓	▓	▓	▓
Design	▓	▓	▓	▓	▓	▓	▓
Implementation	▢			▓	▓		▓
Testing	▢		▢	▓	▓		
Maintenance				▢	▓		

▓ Deals comprehensively

▢ Deals partially

Figure 5.8 Comparative scope of IS development methodologies.

What the comparison shows is that the design methodologies are only part of a wider picture of implementation and none addresses all of the Business Life Cycle. Some methodologies are extensive in scope but others are conspicuously narrow. We can see at a glance the weaknesses in the traditional methodologies and *'code-and-fix'*. They are concerned principally with analysis and design and are weak on the strategic context and on implementation. Conversely ETHICS and

Multiview, while being strong on strategic and analysis and design issues are weak on testing and implementation. Only the RAD methodologies of OLC and ED cover almost the full spectrum of issues. They are not unique in this respect. The point to emphasise here is that both the traditional methodologies and the socio-technic ones, though focusing on particular issues, do not deal with the whole business case of development since either they are not concerned with strategy or with implementation.

In fact a more subtle, but significant message from Figure 5.8 relates to its overriding logic. It is easy to see that the Business Life Cycle, represented by the rows of Figure 5.8 is a concept completely bound up with the waterfall model. There is circularity here which it is difficult to escape from.

Comparative assessments such as this betray a particular mind set which has not escaped the narrowly focused world of methodology. To appreciate this we can turn to an alternative analysis of McConnell (1996) based more on the business management mind-set. He provides a comparative analysis but based on key business management questions.

These key questions are:

- How well are requirements understood?
- How well is the system architecture understood?
- Do I need a highly reliable system?
- How much risk does the project entail?
- Are mid-course corrections likely?
- Do I need to provide visible progress?
- Does the method require high managerial skills?

Figure 5.9 provides a basis for deciding the appropriateness of methodologies covered in this book. In this figure the idea of a cycle is redundant. What matters is avoiding risks and delivering products, meeting targets and people resources. There is, as you might guess, a deeper issue behind this difference in mind sets. This issue is beginning to emerge in business management and should be understood as part of the cultural background to RAD. In a nutshell the issue is 'who runs business?' Is it the IT group or the business manager?

> 'The major issue today is at Chief Executive level, and is about the relationship between IT and the core business.'

> 'How would you describe a journey into the unknown, a place where your compass doesn't work, the sun and the stars are replaced by elusive rainbows, milestones are celebrated and passed but the horizon never gets nearer. The language is strange, the customs new, the pathways intricate and mysterious and you depend on local guides and native bearers. The further you go, the more money they demand to carry on. Gradually the feeling grows that you are going where they want to take you. Does it sound at all like a nightmare? Well, it is — and one that computer managers regularly suffer.'

'I don't know whether you think that management is a science. But, if it is, then it will have its laws. And one of these laws is that you never separate power and responsibility. Never give someone the responsibility for doing something, if you haven't given them the power to achieve it. Even more important, never give anyone the power to do something, if they are not held responsible for the results. Now the introduction of computers to organisations was accompanied by the biggest travesty of this law since the industrial revolution started, in the eighteenth century. Because the users, bless their hearts, still had responsibility for achieving business results. But their power to achieve them was constantly eroded, as they relied, more and more, on computer systems they had neither the time nor the expertise to control.'
Anon, quoted by Grindley, 1993.

CONCLUSION

In this chapter we have explored some of the issues of modern development methodologies based on advances in computing technologies and driven by the business need for rapid development. A major part of the story is the convergence of these methods with socio-technic methods in providing responses to the socio-technic challenge of system development. That challenge is the people problem in organisations which many analysts and commentators have identified as providing the reasons for system failure. Through both types of development methods those problems are being addressed either directly or indirectly.

As always, however, there is a flip side. Addressing or solving one problem opens up another. Our consideration of alternative methodologies during the last two chapters gives us clear messages about where those problems lie. Although ETHICS and Multiview address directly the issues of participation, user-oriented design and human-computer interaction, they provide no method for resolving the conflicting demands from different types of users, from systems developers or from systems themselves. These are the key issues, for what is the point of using socio-technic methods except to resolve them. Yet their solution depends not on the methods but on the personal skill of individuals such as project managers, business managers and users. These personal skills cannot be codified as parts of structured methods.

Similarly the use of OLC, Evolutionary Delivery or RAD methods in general revolves around a series of decisions and actions by key players. Adaptive strategies, which are the cornerstone of all such methods, cannot be executed

Question	Waterfall Model	ETHICS	Multiview	OLC	ED	SSADM	Code-and-Fix
Works with poorly understood requirements	poor	excellent	fair	excellent	fair	fair	poor
Works with poorly understood architecture	poor	excellent	fair	excellent	fair	poor	poor
Produces reliable system	excellent	excellent	good	excellent	excellent	excellent	poor
Manages risks	poor	good	fair	good	excellent	fair	poor
Allows mid-course corrections	poor	fair	fair	excellent	excellent	poor	poor to excellent
Provides visible progress	fair	excellent	excellent	excellent	excellent	fair	poor
Requires little manager skill	fair	poor	poor	poor	poor	poor	excellent

Figure 5. 10 Methodology, strengths and weaknesses. (Source : McConnell, 1996, *Rapid Development*, Redmond: Microsoft Press, Table 7.1. Used with permission.)

according to a rulebook. They are based on individual or collective (agreed and/or negotiated) judgements of when to act, how to act or when to call it a day.

The flip side of both types of method is that personal and interpersonal skills now become the focus of attention. The people resource of organisations moves to centre stage.

SELF-CHECK QUESTIONS

1. What are CASE tools?
2. In what ways is prototyping used?
3. What are the four stages of the 'spiral' model of information systems development?
4. What is prototyping?
5. In what ways can organisations be characterised vis-à-vis IS development?
6. What are the differences in scope of Multiview, ETHICS, Organic Life Cycle and Evolutionary Delivery?
7. What are the key management questions in assessing appropriate methodologies?

WHAT YOU HAVE LEARNT IN THIS CHAPTER

*   *prototyping is a technology led activity which seems to meet many of the needs for GIS development*
*   *modern applications require rapid delivery and new approaches focus on this*
*   *the future for ISDMs is uncertain but it is possible to speculate that people skills and communication will become more significant in systems design*
*   *the real problem with alternative methodologies is to determine which one is appropriate in any situation*
*   *the appropriateness of a methodology must be judged in relation to the business position of the IS being developed and the nature of the project.*

GOING FURTHER

There are a number of texts now available which deal in one way or another with the question of comparative methodologies. An early one which deals with the idea of philosophies and comparative scope of several methodologies is:

AVISON, D. E. and FITZGERALD, G., 1990, *Information Systems Development: Methodologies, Techniques and Tools*, (Oxford: Blackwell).

Its weakness today is the lack of coverage of the more recently developed methodologies.

A text which takes an intelligent and critical look at comparative methodologies and is an entertaining read is:

JAYARATNA, N., 1994, *Understanding and Evaluating Methodologies*. (London: McGraw Hill).

This contains its own methodological framework for comparative analysis and gives an excellent coverage of a few methodologies — worth a look.

Another text which covers similar ground is:

TUDOR, D. J. and TUDOR, I. J., 1995, *Systems Analysis and Design, A Comparison of Structured Methods*, (Oxford: NCC Blackwell).

The general approach is didactic but the book is somewhat difficult to digest because of the complex structure and the lack of clarity in distinguishing methods and methodologies. But the book is valuable as a resource and contains much material — good for reference and borrowing materials from.

If you want to follow up on RAD and get the benefit of up-to-date experience try this text:

McCONNELL, S., 1996, *Rapid Development: Taming Wild Software Schedules*, (Redmond, Washington: Microsoft Press).

This is an entertaining introduction to RAD and the whole set of issues tied up with it. It has a good chapter on comparative life-cycle models.

Chapter Six:
GIS 'Peopleware'

In this chapter we look at the impact which particular types of organisations and particular types of people can have upon GIS projects. We also look at the interaction of GIS and organisations from the other direction — how does GIS alter organisations and alter the working lives of people. We consider:

* *The nature of organisations*
* *The impact of organisational cultures on GIS*
* *The impact of people on GIS*
* *The impact of GIS on organisations*
* *The impact of GIS on people*

'There were no mission statements, no cost benefit analyses, no bench marking, no steering committees, no independent consultants, no progress reviews. In fact there were none of the normal trappings of modern, organisational or systems analysis.

The driving forces during the project were a mixture of chance friendships, the interests, abilities and ambitions of individuals, fortuitous promotions and transfers within the organisation, alliances of self-interest between individuals, hidden agendas, external political forces, the vagaries of funding procedures, the timely maturing of GIS technologies, the advent of companies capable of supplying GIS systems and consultancy, fashion, and finally but most importantly — luck!'
Diploma student.

' "Efficiency means less staff and less staff means a manager has less status" may appear a cynical statement but this is what a manager said to me when I suggested the use of GIS. The more complex the existing system the more work there is for employees and the less chance of anyone being made redundant — including the manager.'
Diploma student.

INTRODUCTION

So far we have approached the problem of introducing GIS into organisations in generalised terms. We have looked at the 'organisational triangle' and at Information Systems Development Methodologies. But, clearly, although organisations display great similarities, each organisation is ultimately unique, having its own culture and its own ways of doing things. An approach that works with one company will not necessarily be appropriate to another. Although the ISDMs provide generalised models describing how Information Systems might be introduced into organisations, each project will have its own unique characteristics.

In addition to understanding generalities, therefore, anybody wishing to introduce GIS into an organisation must also be very sensitive to the particular circumstances created by the nature of the organisation and the people they are working with.

A GIS Commandment:

'GISer know thy organisation'

Constantine (1995) suggests that 'peopleware' represents the third frontier of the computer revolution. In the beginning was the 'hardware crisis'. Everyone thought that the problems associated with Information Systems were due to hardware limitations. Well, the computer manufacturers promptly made available extraordinarily more powerful, more reliable and cheaper hardware but still the problems persisted. Information systems still ran late, over budget and delivered only partial success. So attention soon shifted to what many called the 'software crisis'. If only the software tools available were more sophisticated, software developers would be able to deliver bug-free projects on time. Well, there are now Fourth Generation Languages, structured methods, object orientation and re-usable code. As Constantine says 'Computer-aided software engineering tools sprang up from every point of the compass', but still there are frequent reports of major project failures. The focus of concern, Constantine argues, should now be peopleware, i.e. the behaviour of the people who specify, introduce and, hopefully, ultimately use the systems that Information Technology makes possible. How do real people decide what they need from Information Systems and how, in reality, do they use Information Systems?

Although Constantine was writing about the software industry in general, his argument is clearly particularly appropriate to GIS. We now have reliable, cheap hardware and increasingly sophisticated GIS software. What we now require to make GIS deliver appropriate benefits to host organisations is a realistic focus on GIS peopleware. How do people, and organisations react, when GIS is introduced into their daily working lives? How does GIS alter organisational structures and job descriptions? How can differing company cultures and personalities affect the likelihood of GIS being successfully adopted. In this chapter, we explore some of these GIS peopleware issues.

THE 'POLITICAL' NATURE OF ORGANISATIONS

A necessary starting point for this chapter is to acknowledge the complex, 'political' natures of organisations. In Chapter One, following Campbell and Masser (1995), we observed that there was, unfortunately, little evidence of sophisticated models of organisational behaviour being employed in published accounts of GIS implementations. Indeed, much early GIS literature is extremely simplistic in its view of organisational behaviour. If the problem of organisational acceptance of GIS is considered at all, it seems to be assumed that organisations will behave 'properly' and that individuals within organisations will behave professionally and in accordance with their job specifications rather than according to personal agendas. Using Campbell and Masser's terms, GIS implementation accounts seem to fall into either the 'Technological Deterministic' category, wherein the obvious superiority of new GIS technologies means that organisational acceptance is a forgone, and therefore unconsidered, conclusion, or the 'Managerial Rationalistic' category, wherein any organisational problems consequent upon the introduction of GIS can be easily resolved by a rational restructuring of organisational behaviour. Reality proves somewhat different.

We all know from our everyday lives that organisations are amalgams of groups and individuals each having their own motivations and aspirations, which may, or may not, be consistent with those of their company, but somehow it can seem 'impolite', not quite the right thing, to refer to these realities in published accounts. As Pinto and Azad (1994) observe, however, denying the political nature of organisations does not make them any less political: 'politics are too deeply rooted within organisational operations to be treated as some aberrant form of bacteria or diseased tissue that can be excised from the organisation's body'. Rather than deny the political nature of organisations, it would surely be better to base GIS implementation methodologies upon 'real world' expectations of organisational behaviour than to continue to assume unrealistically rational behaviour.

Pinto and Azad (1994), for instance, suggest that GIS proponents should adopt what they call a 'politically sensible' attitude. They outline three stereotypical attitudes towards organisational politics. The 'Naive' developer believes that organisations run on normative lines, shunning anything that smacks of 'office politics' as morally reprehensible. Such a developer will make unrealistic assumptions and will be constantly frustrated as the organisation fails to behave according to rational expectations. At the other end of the spectrum, Pinto and Azad observe a pattern of behaviour among some GIS implementers who they label as *'sharks'*. Sharks indulge excessively in organisational politics, 'cutting deals' and ignoring official reporting and management structures. Such sharp operators often find, however, that their behaviour becomes counter-productive in the longer term as colleagues learn to suspect their motives. The 'politically sensible' project developer in the middle of the spectrum acknowledges the inevitability of office politics and is prepared to use such politics judiciously to 'ease the wheels' of bureaucracy when appropriate. Politically sensible GIS developers will be aware, for instance, of the role which the 'WIIFM' ('What's in it for me') factor plays in any corporate Information Systems developments and will be aware of the need to balance the divergent interests of all those who may be affected by their projects. Politically sensible GIS implementers will be aware of the differences between 'power hierarchies' and *'power webs'*. Power hierarchies are the formal chains of authority within organisations, as represented in the staffing hierarchy diagrams which most companies draw up. Power webs are the informal, but real, patterns of power which exist within all organisations. Of course, within the webs are *'Spiders'* who are people whose real power exceeds their formal power due to their personalities, status, friendships or specialist knowledge. Pragmatic GIS developers will pay attention to both the official hierarchies and webs of power within the host organisations with which they deal.

Campbell and Masser, similarly argue that it would be more realistic if GIS developers commonly adopted a 'social interactionist' model of organisational behaviour. In the 'social interactionist' model, it cannot be assumed that new technology will automatically be adopted by an organisation, no matter how impressive it may be. Introducing a new system implies a change in behaviour that needs to be *negotiated* between interest groups within organisations. The outcomes of such negotiations will be contingent upon the relative strengths and attitudes of the parties involved.

From this broad foundation that organisations are 'political' and sometimes irrational structures, we will now examine some aspects of the interactions between GIS projects and their host organisations.

THE IMPACT OF CORPORATE CULTURES ON GIS IMPLEMENTATION

Every organisation has its own culture — its own way of doing things. There will be a 'Ford' way of doing things which will differ from the 'Fiat' way of doing things. Often these organisational cultures have formal expressions in books of company rules and regulations. Equally often, there will also be an informal culture built up over the years which might well be at odds with the rule-book — 'Ignore that. This is the way we do things round here'. A successful project leader needs to be a corporate politician with a finely honed sensitivity to the nuances of both the formal and informal cultures of the organisation.

Under the heading of Corporate Culture we have chosen five topics as being most relevant to GIS:

- Corporate or Federal?
- Corporate IS/IT Culture
- Location of the GIS Unit
- Funding model for GIS
- Corporate (In)stability.

Corporate or Federal?

In the GIS literature there have been numerous articles stressing the need to regard GIS as a *corporate* initiative in order to maximise returns on investment. In local authorities and utilities there will be many departments and units which use spatial data, and it is argued that only by involving all such groups can the full benefits of GIS be achieved and the costs of GIS borne. Everyone should use the same OS data. Everyone should use the same property referencing systems. Everyone should share the same road gazetteer, etc. GIS is seen as very much a corporate undertaking. Allowing individual departments to go off and do their own GIS thing will lead to duplication and waste of resource. Korte (1994) for example says: 'A GIS can provide users throughout an organisation access to a common file of spatially referenced data. The strategy integrates and co-ordinates the operations of all users with access to the data. The strategy also offers the greatest opportunity for repeated use of the data. Therefore, making a common GIS database available to the largest possible number of users produces the greatest return on investment.' Levinsohn (1997) similarly argues that 'Many organisations recognise enterprise [i.e. corporate] GIS as the only sustainable approach to widespread GIS implementation... It's not realistic to handle GIS on a function by function basis given the number of functions in large organisations'. Levinsohn also points out that recent technological developments make building corporate GIS more practicable. The common wisdom is summarised in Figure 6.1. Departmental solutions will initially seem cost-efficient but as such solutions multiply within an organisation inefficiencies in terms of duplication of activities, training, data, and maintenance mean that overall costs accelerate rapidly. Corporate solutions have relatively high, and highly visible, start-up costs but as GIS spreads through an organisation, the cost efficiencies of a centrally co-ordinated, corporate approach become apparent.

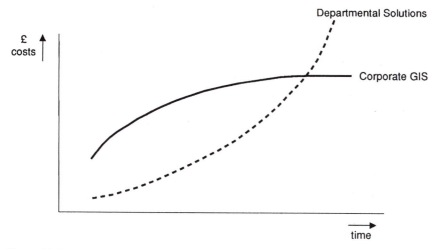

Figure 6.1 Corporate versus departmental solutions.

In the world of real organisations, however, the need for a corporate approach can be pushed too far. Many organisations are in reality conglomerates of essentially separate operations brought together under an umbrella structure for reasons of convenience, historical accident or profit. Think, for example, of the industrial conglomerates which have under their wing a wide range of essentially separate industries and products. Equally, in the public sector, UK local authorities carry out a very disparate range of activities which central government has seen fit over the years to allocate to them. Local authority activities such as Highway Maintenance, Leisure Services, Social Services, Environmental Services, Economic Development etc., have very little in common, each requiring different professional skills, operating under different legislation and having different budgetary requirements. To expect such disparate activities to dove-tail easily into a coherent corporate whole is often unrealistic. Corporate rhetoric may be attractive but, in reality, many large organisations often can probably be more appropriately thought of as consisting of a number of loosely related fiefdoms which sometimes co-operate and sometimes do not, than as tightly integrated unit. 'Turf protection' rather than rational co-operation can be the dominant feature of large organisations. Perhaps a 'Greek Temple' provides a more realistic metaphor for many organisations than does the conventional triangular model. Each pillar in the temple acts as a largely independent body with only a very thin smear of co-ordination at the top of the structure (Figure 6.2).

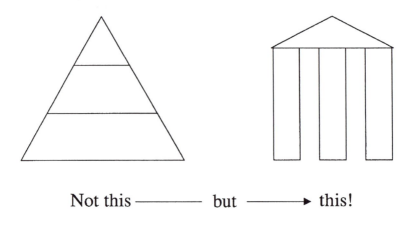

Not this ——— but ——→ this!

Figure 6.2 The 'Greek Temple' model of organisations. (Source: Gill, 1995, *GIS in Powys*, unpublished MSc thesis, University of Salford.)

For GIS the implications of this are clear. Although in a rational world there are strong theoretical arguments which suggest that maximum benefits will be gained from pursuing a corporate GIS strategy, in practice 'politically sensible' GIS project leaders will need to be sensitive to realities of where power and common interests actually lie within their particular organisations and develop programmes accordingly. To try to impose a corporate GIS strategy upon an organisation which pays lip-service to corporate ideals but in which real power rests with service or departmental 'barons', is likely to be a recipe for failure. Better perhaps to forgo corporate ambitions and to build a GIS strategy upon those units within an organisation which display some genuine evidence of shared needs and objectives with regard to spatial data.

Barr (1991) argues that, in fact, a 'federal' approach to GIS development would be more acceptable politically in many organisations. Rather than imposing an inevitably bureaucratic and 'lowest common denominator' corporatist GIS solution upon unwilling departments, Barr suggests allowing natural partnerships to grow between departments with common spatial processing needs. Within local government, for example a 'land development' federation might emerge as might a 'census and socio-economic analysis' group. Such a bottom-up GIS strategy would allow users to retain ownership of their data, choose software appropriate to their particular needs and progress at their own speeds.

It seems that in practice many GIS implementers do, in fact, ignore corporatist rhetoric. Masser and Campbell (1994) describe GIS development in UK local government as being largely a 'decentralised and largely a bottom-up activity'. In some authorities, independent, departmental GIS projects were being simultaneously developed, even whilst multi-departmental systems were being discussed. Similarly the Local Government Management Board (1993) concludes:

> 'Many of the more mature GIS solutions are termed "corporate", but in reality are no more than independent multi-department solutions sharing a digital map base. The concept of sharing of geographic data has not arrived.'

Corporate IS/IT Culture

Organisations vary widely in their attitudes towards computing. Edwards et al. (1991) provide a useful typology of corporate computing cultures:

- *Centrally controlled*: Here, computing 'belongs' to a central IT department. The advantages of this approach are that standards can be maintained, and efficiencies in terms of shared skills and efficient maintenance procedures are available. Historically this was a very common attitude towards IT within companies during the mainframe computing era but became less common as end users became used to the freedom allowed by personal computing. As network computers become more pervasive within companies, a central control philosophy may soon again become common.

- *Free market*: At the opposite extreme to a centrally planned computing culture is a free market approach. Business managers are assumed to be in the best position to know the requirements of their businesses, including IS/IT requirements, and are allowed to meet their requirements from whatever source they believe to be best, whether from the internal IT centre or from external sourcing. In an extreme form this has led many companies to cease to maintain an in-house IT centre. Previously internal IT units have been 'floated' as autonomous, or semi-autonomous, companies to allow them to compete for business from both their former parent companies and external companies.

- *Leading edge*: In some companies there is an eagerness to grasp new technologies. New technology is seen as a competitive tool, and there is a willingness to accept the risks involved in being a pioneer company in order to gain the perceived benefits, and kudos, of being first.

- *Necessary evil:* In contrast to leading edge companies, there are some companies that are inherently conservative in their attitudes towards new technology. In such companies spending on Information Systems will be limited to cases where an unassailable case can be made. Far from seeing benefits in being first with a new technology, some companies can see very strong arguments for being behind the innovators. (The manager of a large IT centre once announced at a conference that he refused to allow his staff to buy any software for which the vendors could not prove that there were already at least fifty satisfactory installations. He said he could not see why his organisation should pay to be an unacknowledged test site for new systems. Admittedly this manager was something of a cynic. He also said that the tenses of verbs in the first release of manuals should often be changed. Where the manual says 'This function does...', it should really say 'This function will ... (once it works properly)'. Anyone who worked with early releases of some GIS products will sympathise with the 'unacknowledged test site' remark.)

For the politically sensible GIS developer, understanding 'corporate IS/IT culture' is important as it provides a clue to what form of GIS development might be acceptable within particular organisations. In a centrally planned environment, proposals for a corporate GIS might be welcomed, and possibly there will be a preference for a single product — 'you can have any GIS you like so long as its ARC\INFO, runs on a Sun Sparc, and conforms to our networking standards'. In a free market culture, there might be more scope for introducing GIS as a series of 'niche' projects, each fulfilling a particular business need. It might be acceptable to recommend introducing different GIS products, each product being justified by its suitability for a particular task.

In those organisations where computing is seen as a necessary evil, a GIS project team can anticipate that there will be a need to produce detailed cost–benefit justifications, and that the financial criteria for approval will be stringent. In a 'leading edge' organisation, the justification for GIS might rest more upon providing attractive demonstrations of potential applications. Executives might be prepared to give approval as an act of faith.

Clearly anyone who wishes to introduce GIS into an organisation needs to be sensitive to the organisation's 'IS/IT' culture and particularly needs at an early stage to determine the nature of the hurdles over which a GIS proposal will have to jump to gain approval. There is no satisfaction in establishing a logical, descriptive case for GIS if an organisation's 'value-for-money' criteria insist that approval will rest ultimately upon there being a cast-iron proof of the financial benefits which will accrue.

Location and Staffing of the GIS Unit

Even modest GIS projects are likely to require additional staff. There may be a need for a GIS manager, applications development staff, support staff, and data technicians. Given the continuing relative scarcity of GIS skills, some of these jobs may need to be advertised at rates higher than may be the norm for the organisation, so that these will be relatively high profile appointments. Furthermore, in many organisations the appointment, grading and location of new staff is a highly contentious matter, and the location of a new unit can provide an occasion for much antler-locking by senior officers. In the context of a GIS project, therefore, the recommendations that the project plan makes about the size and location of any proposed GIS Unit can be politically sensitive and can critically affect the chances of the project being approved. A pragmatic GIS project leader might consider tempering a technically appropriate recommendation with knowledge about what will be politically acceptable.

There are probably three options about where to locate a GIS Unit within an organisation:

- *Computer Services Location*: Where an organisation retains a strong central IS/IT department, it may be appropriate to recommend that this central department develops GIS as a service for user departments. This proposal has the merits that central IS/IT departments are likely to provide professional levels of analysis, design and management. Also, the central IS/IT department is likely to be seen as a neutral, impartial home for GIS by user departments. Initial enthusiasm for GIS, however, often comes from potential users rather than traditional IT professionals and in many organisations the central IS/IT department might be viewed as being unlikely to give sufficient priority to GIS and to be unresponsive to the needs of potential users. GIS does, after all, represent a different kind of computing to the payroll and personnel applications which have been the traditional mainstays of central IS/IT departments.

> 'The IT section generally does not want to get involved with GIS because it involves maps and has nothing to do with "real" computing... If GIS could be placed on a real computer, i.e. UNIX, then it would be taken seriously by IT. The basic fact is that the IT Section are fighting a loosing battle because 90% of users are now PC based.'
> Source: Diploma student.

- *Executive Level Location*: If GIS is being presented as a corporate benefit, locating a GIS Unit within the Chief Executive's Office can be appropriate as it sends a strong message to the rest of the organisation that the GIS initiative has support at the highest level. Also, location in the Chief Exec's Office should provide a neutral location. On the other hand, departmental users may feel that the Chief Exec's office is remote from day-to-day realities and may actually resent such a high profile location. Also some organisations deliberately avoid allocating units such as a GIS Unit to the Chief Exec's Office on the grounds that it is inappropriate for this Office to take on long-term responsibility for service activities.

- *User Department Location*: Locating a GIS Unit within a major user area, for example in a local authority within a Planning or Estates department, can mean that the GIS gains enthusiastic support from immediate users and develops quickly within its initial home. Locating a GIS within a user department can also provide a low profile beginning for the system. Indeed, by allocating some GIS tasks to existing departmental staff it might be possible to minimise the need for additional GIS staff and thus avoid having formally to create a new GIS Unit. Whether such a departmental set-up can subsequently grow into a truly corporate service, however, will depend upon the politics of inter-departmental relations. Also, a departmental GIS might not have the executive level support which it might need to gain widespread acceptance.

Sommers (1994) points out that as GIS projects mature they tend to be 'assimilated' into host organisations and effectively disappear, becoming simply an accepted part of the everyday working lives of the organisations. During their initial stages, however, GIS projects tend to seem novel and contentious. Recommendations about new units, new post, training programmes, using consultants, etc, can therefore be highly sensitive and closely scrutinised.

Clearly the most appropriate location and staffing policies will vary from organisation to organisation, and the task of the GIS project developer is to be sufficiently attuned to the balance of organisational politics to know which proposals will be most acceptable within a particular organisation.

GIS Funding Model

A key issue for a GIS project, obviously, is the nature of the financial support which the host organisation is prepared to provide. The concept of a 'corporate' GIS project might imply that GIS development would be funded from a central, corporate budget. Where such central funding is available, it can provide a stable environment within which a GIS Unit manager can develop a service. In these circumstances, resources might be made available for long-term enhancements of the organisation's spatial database or for application development which otherwise could not be countenanced.

Such central, corporate funding models, however, have become less common in modern corporations. Organisations are increasingly adopting devolved financial planning models in which individual departments or units become responsible for their own budgets and their own profitability. Essentially internal markets are being set up within organisations: if one unit of an organisation wishes to use a service provided by another it will need to 'buy' this service as though it came from an external provider. The notion of a centrally funded GIS Unit sits uneasily within such internally competitive structures. More often the GIS Unit will be expected to prove its value by selling services to client departments. Often client departments will negotiate 'service level' agreements with the GIS Unit for the applications they require. This model of funding means that the GIS manager has less independence to develop long-term GIS strategies, needing always to give priority to the immediate needs of client departments. As Huxhold and Levinsohn (1994) point out, however, this model of funding can result in a GIS service being more finely in tune with the business needs of the host organisation and to a greater degree of commitment from client departments, which having paid for GIS applications from their own budgets, are likely to be committed to their success.

A particular problem with the devolved funding model for GIS developments is that those departments which could benefit most from GIS are not necessarily those which are best able to pay for them. In UK local government, for example, Town Planning departments will often see great potential for GIS within their service, but traditionally planning departments have not had large budgets. Unless the one or more of the large budget departments, such as Housing, Social Services or Transport takes a lead, and is prepared to allow some degree of cross-subsidisation, GIS needs in smaller spending areas might go unrealised.

Corporate (In)Stability

There is often a fundamental incompatibility between the timescales of a GIS project and those of the organisation which is its host. GIS projects have often been long-term developments involving protracted database creation exercises before the system can become fully operative. On the other hand, business managers often require quick results to be able to reassure directors or councillors of the wisdom of their decision to invest in GIS. Also, the business circumstances that made approval of a GIS appropriate may well alter long before the GIS project is complete. A change of management direction, a change in legislation, a change in market conditions, can all mean that a GIS project which was appropriate when it was approved might seem less so now.

Obviously, the degree of instability that a GIS project leader has to confront will vary from organisation to organisation, but a wise GIS leader would always assume instability and plan accordingly. Rather than having a rigid 'grand plan', it is probably better to have long-term goals but to be flexible, indeed opportunistic, about the stages by which goals will be achieved. (The theorists have a grand name for this approach to planning: they call it *'Disjointed Incrementalism'*. One author, however, was honest enough to subtitle his paper 'or the strategy of muddling through'.) The recent emphasis on Rapid Systems Development and on prototyping in the ISDMs discussed in the previous chapter, of course, reflects this concern for flexibility and rapidity in implementation programmes.

The Need for Flexibility

'... the development of conceptual models and system specifications for information systems in an organisation depends on the existence of a relatively stable and clear management structure. This inevitably poses the question of what happens when complex change is a constant factor in an organisation. A structured method will constantly iterate between stages of organisational analysis and conceptual design, never achieving stability for long enough to move to system specification and implementation. This points to the need for methods to be able to produce short term results quickly and for systems themselves to be flexible and loose enough to be able to adapt.'
Source: Diploma student.

Rather than delaying the pay-offs from the system until some far off golden day when all development work is complete, it is wise to plan a series of *milestone* delivery dates when early products from the system can be provided. Early products, such as perhaps a basic mapping service, can do much to sustain support and interest in GIS.

The importance of early deliverables is such that Somers (1994) suggests consciously adopting a 'dual-track' track methodology. Whilst working towards the long-term goal of a full corporate GIS, project teams should also be prepared in the short-term to sacrifice concerns about data accuracy and corporate working in order to deliver immediately useful applications to the organisation: 'Organisations today focus upon the short-term — whether this is good or bad, GIS developers must respond to this environment. Great GIS plans mean nothing to many organisations if there is nothing to show for them today'.

The Importance of Early 'Milestones'

'The installation of the hardware and software was completed on a Wednesday, training in map production was carried out on Thursday and a major map production task involving the creation of over 250 maps for the Head of Estates was commenced on Friday. This was successfully completed in a substantially shorter period and provided maps of far higher quality than would have been possible using manual methods. The system was thus perceived as successful from the first day of installation.'
Source: Diploma student.

Many local government GIS projects seem to target map production as an impressive early milestone in their development.

Although it is important for a GIS developer to be alert to the need to provide early deliverables and to be responsive to ad hoc requests for new applications, it is equally important not to fall into the trap of promising delivery dates which are unrealistic. Acquiescing to unrealistic datelines should be resisted as failing to deliver, or delivering shoddy products, on time rapidly erodes confidence. Business managers will always want results as quickly as possible but they should be educated to understand that pressurising the GIS team into agreeing to unrealistic delivery dates will not result in success.

The Plumber's Approach to GIS Project Management

Never, EVER, tell your bosses anything to do with GIS is simple. Exaggerate delivery times. Suck your teeth. Tell them this is a bad one. Then deliver before just before the deadline. Your bosses will think you are one hell of a manager. Plumbers and car mechanics have been doing this for years.
Anon.

---

or, perhaps, the baker's

'Ask a baker how long it will take to bake a fruit cake and he might say 45 minutes at 350 degrees. But if you are in a hurry and you tell him you've got a blow torch and five minutes, he'll refuse. More project managers should refuse the conditions foisted upon them.'
Kapur, reported in Button, 1996.

THE IMPACT OF PEOPLE ON GIS PROJECTS

> 'There is nothing more difficult to take in hand, more perilous to conduct, or more uncertain in its success than to take the lead in the introduction of a new order of things. For the initiator has the enmity of all who would profit by the preservation of the old system and merely lukewarm defence of those who may profit by the new.'
> Niccolo Machiavelli, 1519.

> 'God created the world in six days, but he didn't have an installed user base.'
> Diploma student.

Most people are inherently conservative. We are comfortable with the familiar and apprehensive of the unknown. On the other hand, within any organisation, there will be some radicals who will be dissatisfied with present circumstances and will welcome change. The success of a GIS project (indeed any project) can depend upon enlisting the support of radicals and assuaging the doubts of conservatives.

A *GIS Champion* is a person for whom promoting GIS within her organisation has become something of a personal crusade. In the brief history of GIS a number of the early landmark developments can be traced back to the persistence and enthusiasm of specific individuals, and it seems likely that behind most successful projects there will be someone with the drive to push, argue and make a nuisance of themselves to get the project done. The characteristics required by a GIS champion have been discussed in a number of papers: Mahoney (1990), for example, writes that a GIS champion 'requires tenacity, single-mindedness, and if all else fails pure logic and not being able to take 'no' for an answer'. Probably the best list of 'champion characteristics' we have seen so far, however, was provided by student at a Diploma workshop.

Steve Gill's List of 'GIS Champion' Characteristics:

doggedly determined, painfully persistent, patient, persuasive, politically aware, flexible, resourceful, imaginative, energetic, calm, calculating, corporate, business-like.

Steve Gill's List of What a 'GIS Champion' Needs:

friends in high places, lead departments, lieutenants, some good applications, some good stories, corporate support, IT support/consent, credibility, an 'image', someone else's CBA, some quick glossy output, an early wicket, an ace up one's sleeve.
Source: Diploma student.

No matter how enthusiastic a GIS champion may be, however, she or he will need to discover *allies*, or in Steve's list, *lieutenants*. Indeed, a key benefit of performing User Needs Study interviews is that it is possible to identify kindred spirits in other departments with whom a supportive network can be formed. To complete the cast list of radicals, it is also vital to have a **GIS Godfather** within the executive level. GIS champions will often be middle-ranking officers who will be seen to have a career interest in pursuing GIS. A Godfather is a senior officer who is above the immediate fray, but who is prepared to promote the project at the highest levels. Because of their longevity and tendency to run over budget and over time, GIS projects can test the patience of Chief Executives. A Godfather at a high level is required to maintain executive support once initial enthusiasms begin to wane: 'Failure to have a champion at executive Level is almost a guarantee for failure or, at best, half-hearted success' (Ives and Crawley, 1995).

On the other side, there will be conservatives who oppose the introduction of GIS into their organisation, either because they do not see the need for it, or feel their positions threatened by it. Many managers feel that their status depends upon the number of staff employed in their group. If a GIS proposal seems likely to reduce staff numbers it might well be opposed, even though it would lead to enhanced efficiencies within the group. Keen (1981) provides a list of 'counter implementation strategies' which such conservatives might adopt:

- **Lay low**: By simply keeping out of the way, avoiding contact with the project team, not replying to memos and being repeatedly unavailable for interviews opponents can hope to frustrate the progress of a project.

- **Rely on inertia**: Assume that without active participation the organisation will continue to work as it has always done.

- **Keep it vague and complex**: If the goals and objectives of a project are kept vague and ambitious, it is likely that will be ample scope for opponents to fight a rear-guard action by asking for endless further clarifications, justifications and revisions.

- **Minimise the implementers' legitimacy and influence**: By holding an implementation team at arm's-length, insisting on formal channels of communication, an implementation team can be made to seem 'outsiders' who do not really understand the inner-workings of an organisation. Denied access to detailed, internal information the project team's proposals can be ridiculed as being inappropriate: 'what do they know, they've only been here five minutes'.

Clearly a GIS project leader will need to identify who are likely to be allies and who may need some persuasion. It is too simplistic, however, to assume that all support will be beneficial, or that all those who raise objections are necessarily irrational malcontents. The naive support of an over-optimist who has read the hype and believed it can in the long-term be an embarrassment. Equally, debates with doubters can be beneficial leading to more tightly reasoned proposals. Opponents can make you reconsider assumptions and can help you avoid making mistakes. Premature agreement can lead to half-baked proposals being progressed. A GIS project leader has to take a mature view of both supporters and opponents and to have the interpersonal skills to negotiate successfully with both groups. The word *stakeholder* has become popular to describe all those groups and individuals who can influence, or be influenced by, a project. Essentially, the GIS project leader needs to identify the 'stakeholders' involved in his or her project, identify their concerns and as far as possible assuage them.

In this chapter we have focused upon the impact which key individuals can have on GIS projects and there is no doubt that many GIS projects do start their life as a consequence of an individual's enthusiasm. Campbell and Masser (1994), however, make the interesting point that the cult of the 'champion' can be carried too far. Although a charismatic and committed individual might be essential to kick-start a project, these very characteristics might make such an individual unsuited to seeing the project through to completion. Once the project becomes a matter of detail and routine, the champion may move on to other new projects. It might be counter-productive in some companies for a GIS project to be identified too closely with a particular individual: 'ownership' of the GIS might better be seen to rest with users or departments, rather than being someone's 'pet project'. Campbell and Masser found that the role of 'champions' had been limited within the sample GIS projects that they studied.

Rather than relying upon an individual to champion the GIS project, many organisations have preferred to give responsibility for developing GIS to groups of individuals organised into 'working' or 'project' groups with members drawn from a variety of departments. Opinions vary considerably about the effectiveness of such working groups. At their worst they may degenerate into little more than bureaucratic 'talking shops' which delay progress or even provide a forum within which departmental antagonisms can be vented.

On the other hand, where the membership contains enthusiasts for GIS, such cross-departmental working groups can provide fertile ground for support networks and innovative ideas to develop. In one Welsh local authority, 'GIS coffee mornings' are regularly held so that the team spirit of users and developers can be maintained. Huxhold and Levinsohn (1994) similarly stress the need to encourage a GIS team-spirit and, intriguingly, they also stress the value of team meals to enhance group cohesion. They use the term *skunkworks* to describe the creation of temporary working environments and project teams, cutting across normal corporate boundaries, which are tasked with introducing new products and systems. Such skunkwork structures can generate strongly focused and motivated teams but, on the other hand, may generate resentment from mainline staff not involved.

Clearly, in any organisation, there will be a unique mixture of personalities, cliques and cabals that will need to be understood and accommodated. The 'politically reasonable' GIS developer will accept that creating a supportive environment among his or her cast list of 'stakeholders' is as important a part of the job as is concern for technical matters.

THE IMPACT OF GIS ON ORGANISATIONS

Information Systems are introduced into organisations to change the way they behave. (If you think about it, the best possible definition of a failed Information System is one which has provoked no change in the organisation into which it was introduced. The whole point of introducing new technology is to change things.)

Information Systems can create and destroy jobs, cause changes in reporting structures, alter decision-making processes and cause shifts in the balance of power within organisations.

It must be expected that any GIS project will have some impact upon the shape of the organisation in which it is based. Groups which previously performed manual drafting tasks might contract, whereas opportunities for new computer-based staff will expand. Depending upon where the GIS Unit is located, responsibilities for processing particular aspects of spatial data might well move from one department to another. Perhaps responsibility for receiving OS data and disseminating it through a local authority might be transferred from its traditional home in, say, the Planning Department to a new GIS Unit located in the Computer Services Department. A department which previously has had to rely upon another department for a mapping service might well find that in future it no longer needs to call on outside help.

Some Realities

'GIS enables the integration of disparate datasets but the consequences of this are far more fundamental than simply enabling senior management to view compilations of data. Changes to peoples' existing practices and status are involved — possibly threats to their continuing employment. This causes problems.

It is likely that the datasets will be maintained (i.e. paid for) by different departments, each having managers and employees with their own agendas. Therefore the full implementation of a GIS will change the existing organisation. If this is not considered and the GIS is not planned as an inherent part of a corporate IS, it is likely to fail. A difficulty is that the effects of a GIS will not be fully understood until the GIS is implemented so they cannot be anticipated — Catch 22.'

Source: Diploma student.

We will use an example to illustrate the organisational issues which introducing a GIS can raise. Imagine that a local authority has developed a corporate GIS which brings together for the first time information from planning, housing, estates, highways, etc., and that in order to maximise the benefits from the GIS in which they have invested so heavily councillors are pressing for an integrated, 'one-stop' enquiry service for residents. Instead of trailing from one local government office to another picking up bits of information, residents will be able to visit one of a number of information points around the town where a comprehensive spatial information service will be available. Consider the turmoil that such a proposal would cause to the 'human' side of organisation. What would happen to the staff presently employed to deal with public enquiries in each departmental office? Who would be responsible for managing the new integrated offices? Would staff in the new information points be paid at a higher grade than staff in the old offices? In order to realise the potential offered by the GIS, the human side of the organisation would have to be considerably changed. The organisational structure would have to be revised, traditional divisions between departments might become blurred, some units might lose staff, some might gain staff, a new unit might emerge, new job descriptions would need to be agreed, new skills introduced.

Against this background it is surprising that as recently as 1996, Musgrave reported that a survey of local government GIS tenders showed that their emphasis almost exclusively remains on hardware, software and database management issues. It seems almost as though some organisations do not really begin seriously to contemplate the changes that will be required to their organisations to accommodate GIS until the boxes are delivered to their doors. In better managed GIS projects, however, considerable preparatory design of the *human* aspects of the GIS system will take place, the aim being to integrate the GIS with appropriate revised office procedures in order to create a coherent, integrated structure which delivers a smoothly functioning and maximally efficient service to the organisation.

When considering the redesign of working practices which introducing a GIS into an organisation might require, an important issue which the GIS developer will need to determine is the degree of *radical* change which an organisation needs and can tolerate.

During the 1990s a strong argument has been made for using Information Technologies to engineer radical change. Under the banner of **Business Process Reengineering (BPR)** management consultants have argued that often in the past Information Systems have been backwards looking, being based on old models of business, and being used essentially to replicate, and thus perpetuate, methods of working which evolved during the manual era. The managers and developers who specify Information Systems requirements it is claimed have been too hidebound by existing work practices to be able fully to envision the new forms of business which modern technology permits. To take the fullest advantages of new Information Systems, the BPR philosophy argues that radically different office procedures need to be introduced. Workflows need to be redesigned to accommodate the new system, rather than the new system being designed to accommodate existing procedures. A new Information System provides a 'clean slate' upon which more efficient workflows and corporate structures can be designed.

Business Process Reengineering

'Reengineering is the fundamental rethinking and radical redesign of business processes to achieve dramatic improvements in critical contemporary measures of performance, such as cost, quality, service and speed.'
Hammer and Champy, 1993.

'Business reengineering isn't about fixing anything. Business reengineering means starting all over, starting from scratch.'
Hammer and Champy, 1993.

'The computer industry has been building the wrong systems. They have sought to automate procedures that already exist but that negate the true value of the technology. They should have been thinking about reinventing the business.'
Martin, 1997.

Clearly the BPR concept has some relevance for major GIS projects. In many organisations, map-handling systems have evolved over decades and have become engrained into the cultures of the organisations. Adopting a BPR philosophy, however, an implementation team might decide that introducing a corporate GIS presents an opportunity to make a clean break with old methods and decide radically to redesign workflows to maximise the benefits available from the new technology. A number of papers have indeed recommended that GIS implementers should indeed adopt the BPR approach. Aybet (1996), for example, argues that the low success rate so far achieved by UK local government GIS projects suggests that they are not being implemented in the correct manner and suggests that greater success would be achieved if such implementations were viewed within a business re-engineering perspective. Mingins (1996) similarly argues that regional electricity companies should regard GIS as a central element in the reengineering of their core business processes.

GIS *is* a Radical Technology

'Remember — implementation of a GIS means the introduction of the biggest company cultural change since your accountants changed from using quill pens and ledger books to computers, over many years using incremental technology changes. You are going to move through all the accountant's data processing steps and beyond, *in one go*.'
Ives M. J. and Crawley K. J. 1996.

As with many good ideas, however, the BPR can be pushed too far. Although BPR continues to be very fashionable, there has been something of a backlash against some aspects of the implementation of these ideas. Often BPR has seemed to be associated with 'down-sizing', 'right-sizing' and whatever other euphemism is used for reducing staffing levels. The BPR philosophy seems to adopt a very normative, technocratic view of how modern organisations should behave, neglecting the human realities of change. Indeed, the very use of the word 'engineering' within the title of BPR will rise the hackles of some critics. The suggestion that systems which involve human beings can be 're-engineered' in a manner analogous to mechanical systems is, of course, anathema to those who believe in a *socio*-technical approach to change: 'The rock reengineering has foundered on is simple — people. Reengineering treated people as if they were just so many bits and bytes — interchangeable parts to be reengineered. But no one wants to be re-engineered. BPR became a code word for "mindless corporate bloodshed." ' (Peltu, 1996).

A BPR 'Joke'

In an executive washroom the other day I overheard this exchange:
    'Hey, how's it going? I haven't seen you in a while.'
    'I got re-engineered.'
    'Hey, too bad'.
Adams, S., 1996.

With regard specifically to GIS projects, Chrisman (1987) provided an early warning against too readily riding rough-shod over apparently irrational practices: the duplications of data holdings which might from the analyst's point of view seem ripe for rationalisation may have deep-seated human and organisational justifications. Concerns for job security, worsening conditions of service, etc., engendered by radical change might cause morale to wilt at the very time that enthusiasm is required to ensure a new system is accepted. A GIS development team, with executive level guidance, will need to judge the degree and rate of change that an organisation can assimilate. For a GIS to be successful, the staff involved must be committed to making it so. If staff feel threatened by the speed and severity of the changes they are experiencing, they may be obliged to accept them but, as Huxhold and Levinsohn (1994) point out, 'compliance' is not the same as 'commitment'. Rather than adopting a radical, BPR style approach to GIS implementations, it might sometimes be more in keeping with the ethos and realities of an organisation to adopt an evolutionary style, introducing GIS gradually and allowing workers to realise the benefits at an acceptable pace (Wiley, 1997).

Regardless of the pace and degree of organisational change provoked by individual Information Systems projects within particular organisations, it is clear that the *cumulative* impact of an ever-increasing dependence on Information Technologies is causing the form of organisations to change dramatically. Indeed, it is clear that the relevance of the familiar triangular model is being rapidly undermined. This model was, after all, developed prior to the widespread introduction of Information Systems at a time when organisations did require large numbers of clerks and middle ranking officers to sift and process data and even larger numbers of operational level staff. The triangular model is based upon an essentially *hierarchical* view of organisations but introducing Information Systems tends to blur previously rigidly hierarchical reporting structures. With paper based systems memos and reports tend to be passed up and down the hierarchy of command and to be sent only to restricted distribution lists, but in companies which have invested in Information Systems, information can be much more widely available. Instead of waiting for middle managers to prepare paper reports, senior officers can, if they wish, tap straight into corporate databases. In companies which have adopted electronic mail executives can dip into the electronic mailbag at any level, and conversely low level officers can easily bypass normal procedures and send messages to everyone on the system. To some degree, information flows are moving from 'hierarchy' to 'ad-hocracy'.

In an early review of the impact of urban data systems on USA local governments, Downs (1967) found evidence to support this pattern of less rigidly hierarchical structures. He found that information systems could allow top-level officers direct access to information which previously would have been 'screened' by intermediate officers. As a result middle-ranking officers lost power to higher level officers and politicians. Indeed, the need for middle-ranking officers was reduced and the job descriptions of those that remain tended to be redefined to encompass new tasks. In general, there has been a trend in recent years for 'flatter' organisational structures in which middle layers of management are removed. The rather ugly word *'disintermediation'* has been coined to describe the process by which intermediaries can be bypassed by the use of Information Technology. The triangular diagrams in future should clearly be drawn much more shallowly.

Commentating on emerging trends some authors envisage even more radical changes in the shape of organisations. Handy (1989), for example, argues that a 'shamrock' model rather than a triangular one might more accurately describe the structure of organisations in future. One leaf of the shamrock will be a small group of 'core' workers. These will be the key executive staff which an organisation needs to manage its business. These core staff will be highly paid and tied to the organisation by long contracts and attractive benefits. The second leaf will be the 'contractual' fringe, which will consist of the lawyers, accountants, computer specialists, etc., who will be called in on a contract basis to perform specific tasks. Whereas in the middle layer of the triangular model it is expected that each company would have its own accounts, personnel and computing departments, in the emerging shamrock model these activities are likely to be contracted out, thus reducing overheads and increasing flexibility for the hiring organisation. The third leaf of the shamrock will be a 'flexible' (read part-time, temporary, non-unionised) workforce which will be hired and fired as demand requires.

Handy makes clear that this shamrock model, in which only the core workers have conventional careers inside the organisation, is made possible by Information Technology. IT facilitates the spatial disaggregation of activities that previously had to be located together for purposes of control and co-ordination. Why pay expensive computing staff in the UK, when there are lots of highly efficient, cheap programmers in Bombay with whom your firm can easily communicate by satellite? Why pay British rates for word-processing when operators in the Philippines can turn round the work just as quickly but much cheaper? Why incur the office overheads of employing your own accountants when you can communicate, via e-mail, fax and video-telephone, to an accountant anywhere in the world just as easily as if the accountant were in the next office? Indeed, even the core workers may only turn up in the office occasionally as an increasing number of them choose to telecommute.

It is interesting to speculate about the role of GIS in the shamrock organisations of the future. Once a local authority has all its planning applications data on a GIS system, why should development control officers continue to work from central council offices? Perhaps they could work from home tapping into the GIS over a link? Perhaps, indeed, planners will soon be able to take their departmental databases 'on-site' with them via hand-held devices? Indeed, many utility companies are already experimenting with mobile GIS/GPS systems in order to maximise the time their field operatives actually spend away from their offices. Possibly GIS activities might be better allocated to the 'contractual fringe' in future. One wonders whether organisations actually do need their own GIS Units and their own GIS databases or whether it might be more efficient for organisations to contract out responsibility for maintaining spatial databases to specialist 'GIS contractors'? There is certainly also scope for the 'globalisation' of some aspects of GIS activity. Already there are 'digitising shops' in the Third World which can complete data conversion exercises at much lower prices than western agencies, using satellite links to return data promptly to western clients.

THE IMPACT OF GIS ON PEOPLE

'The proper use of technology is the humane use of humans.'
Norbert Weiner.

Surprisingly little seems to have been written so far upon the impact of GIS on the job content and job contentment of 'line' workers within organisations. Alter (1992), however, provides an intriguing list of the characteristics of a good job which we can use as a framework for discussion:

- **Skills**: You are able to use and increase your skills.
- **Meaningfulness**: You feel your job contributes significantly to the purpose of your organisation.
- **Autonomy**: You feel you have reasonable discretion in the way you organise your work. You do not feel under excessive supervision.
- **Social relations**: Your job allows work to be an enjoyable social experience. You can collaborate and communicate with colleagues.
- **Psychological demands**: Your job is challenging and engaging but not excessively so. There is a mixture of novel and routine tasks. You do not feel permanently stressed.
- **Balance with life beyond work**: Your job does not demand so much of your energy and time that it interferes excessively with your ability to participate in family and community life.

A feature of the effect of Information Technology upon jobs is that it is *indeterminate* in that so much depends upon the particular circumstances in which it is introduced. The same system could have widely differing impacts upon employees depending upon the circumstances of its introduction — the attitudes of management, the job descriptions provided for operatives, the physical arrangements of the office, level of training and support provided etc. We will use Alter's criteria to look briefly at some of the effects which Information Systems have been observed to have upon job experiences.

If we look first at **skills**, Information Systems can often pose a threat to traditional skills. The term *deskilling* is used to describe the process by which introducing an Information System removes the need for traditional skills. With GIS, there is a very obvious likelihood that GIS will dramatically reduce the demand for traditional manual drafting skills. The cartographer who drew most of the diagrams in this book tells us that for the first week of his professional training he was allowed to do nothing other than print, and reprint, the alphabet until he developed a pleasing free-hand printing style. He then progressed to skills such as applying colour washes, using stencils, line shading, etc. With the adoption of Computer Cartography and GIS many of these hard won and rightly cherished skills have become less relevant — the machine produces the text, colours and shadings. Many traditionally trained cartographers and draughtspeople, Terrier clerks and building control technicians, might well be forgiven for regarding the advent of GIS as devaluing their traditional, hard won, skills.

On the other hand, many people are intrigued by new technology and look forward to gaining new technological skills. Here the appropriate term is not deskilling but ***reskilling***. In one GIS feasibility study with which one of the authors was associated, two Terrier clerks whose responsibility it was to draw property ownership boundaries onto OS map sheets were initially very apprehensive about the proposal to introduce GIS and extremely diffident about their ability to transfer to the new technology. When the situation was explored with them, however, it became apparent that the clerks had considerable knowledge of the Terrier process, built up over many years, and that this knowledge would remain essential even if a GIS was introduced. The GIS would remove from them the manual drafting tasks but would not devalue their core knowledge. When this was understood, the Terrier clerks became enthusiastic to learn the new skills associated with GIS as they regarded it as a means of enhancing their abilities to do their jobs. Whether a GIS is regarded as leading to deskilling or reskilling often seems to be largely a function of the personalities involved and the degree of sensitivity and support with which changes are introduced.

Alter refers to task variety and task scope as issues that impact upon the **meaningfulness** of a job. ***Task variety*** refers to the range of different things people are asked to do in their jobs. Generally the greater the variety of tasks the better a job is considered to be. Sometimes an Information System, by automating a repetitive task, can release a person to take on other tasks and thus expand task variety. If, for example, a GIS reduces the amount of time cartographers have to spend on the manual tasks of map drawing, this might mean that cartographers can spend more time on the creative tasks of map design or even become involved in new tasks. Conversely, Information Systems can sometimes be used as a means of increasing job specialisation and thus reducing task variety. ***Task scope*** refers to the size of the task compared to the overall purpose of the organisation. Generally, highly constrained jobs, e.g. fitting the wheel nuts on the rear left wheel on a car assembly line, are considered unattractive and people usually prefer to have jobs with larger scope. Interestingly GIS has already had the effect of enlarging the task scope of some cartographers and draughtspeople. In the old days the role of a cartographer was usually constrained purely to producing a final map — the analysis and data-processing would be done before the cartographer became involved and equally the use to which the map was put after it had been drawn was not the concern of the cartographer. Today, however, because GIS software contains analytical routines as well as display functions, and because less time needs to be devoted to mechanical drafting tasks, some cartographers have become involved in the earlier research and data processing aspects of projects.

A particularly important consideration for some GIS operatives with regard to the meaningfulness of their jobs is the *increased abstraction* of their tasks when they move from manual to digital mapping. Many cartographers, planning technicians, draughtspeople, etc. will have been attracted to their jobs because they actually like manipulating maps, paper, inks, tee-squares and drawing boards. Crafting a well-designed paper map can give a satisfaction similar to that derived by an artist creating a painting. With GIS, however, this direct, tactile contact with physical mapping is lost as GIS operatives work, not with real maps, but with *representations* of maps on VDU screens. Map work has become ***computer-mediated*** in the sense that computers are placed between operatives and the objects of their tasks. The working environment of GIS operatives is more like that of a computer programmer than that of a traditional drawing board jockey. Whether traditionally trained cartographers would have chosen to become cartographers if they had known their working lives were going to be spent slaving over a hot Apple Mac screen is a moot point.

Information Systems can either increase or decrease the **autonomy** people feel in their jobs. For researchers and policy analysts GIS systems on PCs and workstations can greatly enhance their freedom to experiment. In the old days a planner wanting to run, say, a Spatial Interaction Model would have had to negotiate with the owners of the local authority's computer (usually Treasurer's department who would only allow foreigners to use 'their' machine on Fridays at 4.00pm when they'd finished doing important things like the payroll). The planner would also have had to have got a boss to agree to pay for the computer resources used. Because of the bureaucracy and cost involved, running a computer model was a big deal and if the model did not prove useful everyone would know about it. Today, however, a planner can quietly experiment upon her desk-top GIS. If the model works — fine, if it does not, then it's no big deal because nobody else need know about it. Increased data availability, cheap computer power and suites of analytical and mapping routines greatly enhances the autonomy of researchers and policy analysts. The term ***empowerment*** is used to express this ability of Information Systems to allow some workers to take fuller, and more independent, control of their working environments.

On the other hand, for many employees, Information Systems have been used to increase levels of supervision and to reduce autonomy. In conventional computer data processing systems, for example, data entry modules might be set up to monitor how many keystrokes each operative achieves, how many incorrect keystrokes, how many breaks an operative takes each day, etc. The digitising task in GIS can similarly be controlled — how many polygons did this operative achieve today, how many of them compiled first time? Nor is it only at the operative level that Information Systems can be used to monitor performance. As has already been mentioned the advent of powerful information and electronic communication systems allows executive-level officers to have greater access to information and this can mean that middle-level officers can feel that their scope for action is diminished and that their performance is being more closely scrutinised.

Information Systems and Ethics

As Alter points out issues such as using computers to monitor staff and to pace work raise concerns about corporate morality and ethics. If a company is going to use Information Technology to monitor an employee's performance, is the company morally obliged to tell the employee? To what extent are employees entitled to a degree of privacy in the performance of their tasks?

Many of us regard the **social relations** of work to be important. Meeting people to discuss work-related issues (and the more important matters in life such as how badly Trinity played last Sunday) is an important part of our work experiences. In a conventional Town Planning Office, for example, wandering down to the basement to unearth a planning application, nipping up to see the Highways lads, visiting the typing pool to attempt to persuade them to type letters before they become totally irrelevant, maintaining good relations with the technicians, etc. were all welcome parts of the job. Once a GIS is introduced, however, there should be much less need for planners to leave their desks. Having all the information necessary to do your job available on your desk-top might be great for productivity, but it might not do much for social relations at work.

Has Anybody Asked Them?

The diagrams in this book were drawn by Steve Pratt, the Huddersfield University cartographer, using Adobe Illustrator on an Apple Mac. Within the last five years Steve's job has changed almost completely. He has gone from a traditionally trained and skilled manual cartographer to a skilled and enthusiastic Apple Mac designer. Steve only uses his drawing board now for the few jobs which he feels can still be done better by hand.

Fortunately Steve had the confidence and enthusiasm to welcome the new technology and was prepared to put in the effort required to learn the new skills. But perhaps not everyone has Steve's outlook and ability. The rhyme at the beginning of this book points out that the success of Information Systems ultimately lies in the hands of the 'Joses, Ahmeds and Kates'. The success of GIS systems lies in the willingness of people like Steve being prepared radically to transform their working lives.

Surprisingly, however, in the GIS literature we cannot find much evidence of any systematic enquiry into the impact of GIS on the work experiences of operative staff. Very little seems to be written from the viewpoint of the people at the sharp end of GIS whose traditional skills are challenged by GIS. Does GIS, on balance, increase the quality of peoples' working lives as measured along Alter's criteria? Are draughtspeople generally happier now they are using GIS than they were in the old manual days? Do they get a boost from using new technologies? Or, do they feel greater pressures to be more productive now they have GIS? Do they feel that they have been deskilled or reskilled? Nobody seems to know — perhaps because nobody has yet asked?

CONCLUSION

> 'A good team can deliver a successful project even if the technology is deficient. However, if the team is poorly managed the project is unlikely to be a success, even if blessed with best of breed technology.'
> Coote, 1997.

In this chapter we have looked at the interactions between GIS and organisations and people. We have argued that GIS project leaders need to be sensitive to the nuances of the organisations, in which they work and need to be flexible in their approaches to people. People are the key to GIS success.

We will finish, however, with the ironic point that no matter how carefully and sensitively you plan the introduction of your GIS, no matter how charismatic your project leaders and supportive your managers, the outcome will almost certainly *not* be that which you anticipate. Empirical studies show that with most projects the most significant impacts of Information Systems are accidental rather than planned (Robey, 1987). Information Systems tend to have ramifications well beyond those departments and employees who will be most immediately and obviously affected. Unforeseen problems and opportunities will knock you off your planned development path. No matter how carefully you plan, you almost certainly will not end up where you planned to be!

SELF-CHECK QUESTIONS

1. Attempt your own definitions of the following :
    — Peopleware, Technological Determinism, Managerial Rationalism, Social Interactionism.
2. What are the characteristics of the following attitudes towards IS/IT spending within companies:
    — Central control, Free Market, Leading Edge, Necessary Evil
    — How might these differing attitudes affect GIS developments?
3. How might different budgeting arrangements within companies affect GIS planning within companies?
4. What are the characteristics of a GIS 'champion'? Why might reliance on a GIS champion be counter-productive in some circumstances?
5. What is meant by 'stakeholder' analysis?
6. What is a 'skunkworks' team?
7. Attempt to define the meaning of 'Business Process Reengineering' and consider its strengths and weaknesses as a model for planning the integration of GIS into an organisation.
8. Why might the traditional triangular diagram of organisational structure be losing its relevance?
9. What is the meaning of the following terms:
    — deskilling, task variety, computer-mediated, empowerment?

WHAT YOU HAVE LEARNT IN THIS CHAPTER

*   *Organisations have different attitudes towards IS/IT*
        — *'centralised', 'free market', 'leading edge', 'necessary evil'.*
*   *There are key people who affect the likelihood of GIS success*
        — *'champions', 'Godfathers', 'lieutenants', 'opponents', etc.*
*   *The 'human' side of GIS implementation needs conscious planning*
        — *Business Process Reengineering or Evolutionary.*
*   *IS/IT is altering the shape of organisations*
        — *from 'triangle' to 'shamrock'.*

* *IS/IT is altering the work experience of many employees*
  — 'deskilling', 'reskilling', 'empowerment', 'computer mediation', etc.
* *Significant impacts will often be unanticipated.*

GOING FURTHER

An excellent text about the organisational and human aspects of introducing Information Systems into organisations is provided by:

EASON, K., 1988, *Information Technology and Organisational Change*, (London: Taylor and Francis).

For anyone interested in what the future may bring in terms of our working lives, any one of Handy's books will make stimulating, and easy, reading. The 'Shamrock model' is described in:

HANDY, C., 1989, *The Age of Unreason*, (London: Hutchinson).

Handy also describes the 'Federal Organisation' and the 'Triple I Organisation' models in this same book.

A more recent book about the future of organisations, which particularly stresses the role of IT, is:

MARTIN, J., 1996, *Cybercorps: The New Business Revolution*, (London: McGraw-Hill).

There is a large literature on Business Process Reengineering, but a good starting point remains an original text which was largely responsible for the booming popularity of BPR among managers:

HAMMER, M. and CHAMPY, J., 1993, *Reengineering the Corporation: A Manifesto for Business Revolution*, (London: Nicholas Brearley).

Anyone who really wants to understand behaviour inside modern corporations should make a habit of reading the 'Dilbert' cartoons. These are very revealing, and much more fun, than management textbooks. The book below provides, amongst other things, Dilbert's views about Business Process Reengineering:

ADAMS, S., 1996, *The Dilbert Principle: A Cubicle's Eye View of Bosses, Meetings, Management Fads and other Workplace Afflictions*, (London: Boxtree).

Chapter Seven:
The Way Ahead

In this final chapter we look ahead to the future of GIS within organisations. We do this by offering some final thoughts about the two major themes which have run through this book:

> * *how 'special' is GIS compared to conventional IS systems?*
> * *what should be the role of methodology within the GIS implementation process?*

'GIS used to be a boutique industry, but now it's becoming mainstream. Anyone who doesn't keep up with this development won't last long.'
Kim K., President of Harvard Mapping and Design, quoted in Wilson J. D., 1997.

INTRODUCTION

Two issues have dominated the earlier chapters. The first is whether there is anything special about GIS? Is the development of GIS really just one of mainstream IS history on fast forward or are there aspects of GIS which are sufficiently distinctive or problematic to make it worthwhile continuing to regard GIS as a specialist area of interest? Should companies regard GIS as special projects, or would they be better advised to regard GIS as simply another computer system to be evaluated and implemented according to standard procedures? The second issue is about the role there is for the *conscious* use of Information Systems Development Methodologies within GIS projects. Would GIS benefit being dealt with in standard ways? In a rapidly evolving world can we identify an appropriate methodology?

In this final chapter, we take up both these themes again in the context of current and likely future developments.

WHAT'S SO SPECIAL ABOUT GIS?

Conventionally, GIS has been regarded as a specialist sub-area of activity, distinct from the mainstream of Information System development. GIS has had its own literature and its own journals. It has had its own specialist vendors: GIS software has been supplied so far by niche vendors, such as ESRI, Intergraph and Smallworld rather than mainstream companies such as Oracle or Microsoft. Within universities GIS has been taught primarily by geographers, rather than by business information systems specialists. In previous chapters, we have remarked how little impact ideas from the dominant methodologies in mainstream information systems development seem to have had upon GIS projects. GIS has formed its own community and interaction with the broader sphere of Information Systems seems not always to have been strong.

It can be questioned, however, whether this quasi-separate status for GIS is either sustainable or desirable. The convergence of computing towards open systems and interoperability means that the hardware and software justification for a separate status for GIS is disappearing. As the quote from Kim at the top of this chapter makes clear, vendors which try to stand aside from mainstream trends simply will not survive. Indeed, commercial realities are already forcing GIS vendor companies to look for alliances with mainstream companies: increasingly we may in future look to the likes of Oracle and Informix for GIS solutions. Most importantly in the context of this book, the insistence of companies that their investments in GIS systems must show an acceptable rate of benefit will mean that increasingly mainstream evaluation and development methodologies will be applied to GIS projects.

We will consider further the convergence of GIS with the mainstream Information Systems industry under three headings:

- Technological convergence
- Commercial convergence
- Organisational convergence.

Technological Convergence: When the first generation of GIS software was being written, the dominant model upon which mainstream database software was based was the *Relational model*. In essence, this model imposes a stunningly simple structure upon the ways in which data are held in that *all* data are regarded as being held in conventional two-dimensional tabular formats. This simple, elegant model has proved to be exceptionally successful for most corporate information systems applications. Indeed, all the currently dominant commercial database packages — Oracle, Ingres, Sybase, Informix, Access, etc. — can to a large degree be regarded as relational packages.

When the authors of GIS software examined the suitability of the relational model for handling geographical data, however, many of them concluded that it had significant limitations. The relational model functions best with relatively simple datasets wherein a change in one value is not likely to impact greatly upon other values, but geographical data tends to be highly interconnected and complexly structured. The relational model requires data to be of fixed lengths but in geographical datasets features such as roads or rivers are of varying lengths. Relational databases tend to be optimised for short data recalls but geographical inquiries usually require large amounts of data to be recalled and displayed. Relational databases are optimised for very short editing cycles, but planners working on spatial data for a section of a city might need to take months to complete an update. Relational databases are optimised for linear searches, whereas spatial data interrogations will usually require two, or three, dimensional searches. Simply put, many GIS software vendors concluded that there was a very significant mismatch between the complex nature of geographical data and the rather simple model of the dominant mainstream database model of the day.

Twenty years ago, therefore, there was a basis for arguing for a technological distinctiveness for GIS: The then dominant database model simply did not 'fit' geographical data. Accordingly, many vendors adopted what became known as the 'dual architecture' model for GIS software. Rather than attempting to cram the complexities of geographical data and geographical queries into the inadequate relational model, they wrote their own bespoke software to handle specifically 'geographical' data. This software could be optimised to store and manipulate geographical features and could provide the spatial search algorithms likely to be required by users. Relational database software was incorporated into the dual model only as a convenient vessel for storing attribute data, i.e. data describing the condition of the spatial features. This dual architecture is embedded in the name of the 'ArcINFO' software which remains a dominant product today. 'Arc' is specialist software written by the ESRI company to handle, interrogate and display spatial features whereas INFO is an early implementation of the relational model which ESRI 'bought in' to handle attribute data.

A number of technological developments, however, have weakened the case for there being a continuing technological basis for a distinct GIS software industry. Within the relational database world, the major vendors have been increasingly prepared to sacrifice the purity of their adherence to the relational model in order to be able to incorporate complex datasets, including geographical data. Most of the major relational database vendors now offer 'spatial' extensions to their core products that offer spatial indexing and enquiry facilities. Indeed, the most recently agreed standard for the Structured Query Language (SQL), which is the core language of the relational model, has had spatial operators built into it. One real possibility is that the mainstream database products will engulf traditional niche GIS products. In this respect it is interesting to note that Oracle, and Microsoft, are principal partners in the Open GIS Consortium and are therefore contributing on an equal basis with the traditional GIS vendors towards plans for the future of the GIS industry.

A further technological development which points towards technological convergence is the growing maturity of object-oriented database technology. Whereas the relational model is in many ways unsuitable for geographical data, the Object-Oriented model might almost have been designed for geographical data. The O-O model is at its best when handling complex, interactive data structures. As the O-O model gains an ever-larger acceptance within the mainstream information systems industry, unfamiliarity with the types of computing challenges offered by geographical datasets will generally diminish. Riding partly upon the back of the O-O model, the increasing emphasis upon interoperability and open systems computing also strongly argues for technological convergence between GIS and the mainstream industry. Rather than exporting data from a conventional database into a specialist GIS for spatial processing, 'components' of GIS functionality increasingly will be incorporated into mainstream applications. GIS will be used less and less as separate packages and more and more will be embedded within other applications. GIS will be only one icon on a crowded screen, or more likely still, will be represented as one pull-down menu within the familiar layouts of Windows applications. The incorporation of a limited GIS functionality taken from the MapInfo company into the Microsoft Excel spreadsheet package is merely a first straw in the wind.

Commercial Convergence: The collaboration between MapInfo and Microsoft with the Excel package is also just one, very visible, example of a commercial convergence which is occurring among GIS companies and between GIS companies and mainstream software vendors. Increasingly previously fiercely competitive companies are having to collaborate and establish strategic alliances with each other, so that the distinctiveness, indeed independence, of traditional GIS companies may become blurred. A criss-cross network of alliances is emerging between GIS companies and mainstream vendors (Wilson, 1997). Microsoft, for example, not only has links with MapInfo but also a collaboration with ESRI. Oracle has links with most of the major GIS vendors. We are probably witnessing a shake-out of the boutique industry.

This commercial convergence is being partly encouraged by technological imperatives. The trend towards interoperability and open systems computing mean that software companies have to be more open towards each other and more willing to share technologies in order for their products to be compatible.

There are also some hard-edged commercial realities driving this commercial convergence. The rates of technology change and version upgrades are now so rapid that the relatively small GIS companies will have a difficult time keeping pace independently. They need access to the technologies being generated by the mainstream companies:

> 'Over the past five years we've come to the realisation that we can't do everything. We need to be able to plug into existing information systems. This will guide the industry for a long time. Its what the users want....
>
> Application upgrades are hitting users at a staggering pace. If integration is to be maintained, GIS developers must keep abreast of each new upgrade. A company that fails to integrate its system close to the leading edge is a company that easily can be left behind.'
>
> Boyle, V.P. Intergraph, quoted in Wilson, 1997.

So, it seems that the distinctiveness of GIS at a commercial level is lessening. In future, organisations may end up using GIS functionality which originates from a specialist company but they will not necessarily know this to be the case as the software will have been 're-badged' and sold on as part of a 'total solution' by a mainstream vendor.

Organisational Convergence: As discussed in earlier chapters, one of the atypical aspects of many GIS projects has been that they have often been championed, and 'owned', by non-technical people. Typically middle managers or professionals such as town planners or highways engineers, have seen GIS as a way of removing a bottleneck within their operations and have initiated GIS projects. Consequently, one of the suspected reasons for the failure of some GIS projects within organisations has been that because of these departmental origins, the projects have not been integrated fully into organisational information strategies and have not benefited fully from the conventional scrutinies which are applied to other information systems projects.

Some authors have argued that GIS projects would be more successful if their champions were less concerned with the 'G' and concentrated more on the 'IS'. Waters (1992), for example, argues that the very concept of a 'GIS consultant' is flawed. What companies really require are traditional Information Systems consultants, well versed in the conventional skills needed to bring complex, large, projects to fruition, who also, as a subsidiary skill, understand geography. Similarly Musgrave (1996) in a paper interestingly entitled 'Ignore GIS — information management is the key' argues that organisations should focus first and primarily upon devising appropriate corporate information management strategies and evaluate GIS only once this overall context has been established.

In the past GIS may sometimes have been seen as 'maverick' projects existing outside the normal procedures of information technology planning. In future this distinctiveness may lessen as organisations incorporate GIS projects more fully into mainstream Information Systems strategy planning cycles. If this occurs there should be a greater explicit use of Information Systems Development methodologies in future GIS projects. Because corporate GIS projects are large, complex undertakings that cut across many vested interests, they will remain difficult to achieve and will continue to provoke the human problems discussed in Chapter Five. Increasing, explicit use of mainstream socio-technical methodologies, however, might lessen the traumas associated with GIS implementations.

But 'Geography' Remains!

'Owning a spreadsheet doesn't make you an accountant!'

So far, we have argued that GIS will become a less distinctive area of activity in the future. Convergence in computer technologies, commercial pressures and increasing corporate scrutiny will lead to GIS being integrated into the mainstream of corporate information technology.

On the other hand, the very availability of GIS functionality across corporate desk-tops will put into higher relief an issue which technology cannot address. Many employees, at all levels, who previously have not had exposure to maps, spatial data and spatial enquiry methods will suddenly be able to click on buttons and produce mapped outputs. This raises two questions. First, will users from traditionally non-spatial disciplines actually appreciate the value of analysing and presenting their data spatially? Will the value of 'spatial thinking' be apparent to them? Second, if they do become enthusiasts will they appreciate that there are pitfalls in spatial analysis? Practitioners from the spatial disciplines such as Town Planning, Geology and Geography have long realised that 'spatial data' is tricky stuff which can lead the unaware into making wrong assumptions and interpretations. We know about scale effects, boundary effects, map projection effects, spatial autocorrelation, the subjectivity and subtleties of map design and the like, but is it reasonable to expect a marketing executive or a business manager to have this knowledge? After all, we all have spreadsheets on our machines these days but that does not mean that we can all claim to be accountants.

Perhaps for GIS educators there needs to be a shift of emphasis. We will need to spend less time explaining the technologies of GIS but more time explaining why spatial analysis is useful in business and how spatial analysis should properly be done. Increasingly GIS will be an unremarkable button on the screen. Understanding when to click on that button and how to interpret the results produced will be the key skills which educators are required to provide.

> 'Maps are not simple graphic displays of information. An experienced worker can show great skill in using maps. These skills of design, analysis and interpretation are a real craft not accessible simply by looking at maps or producing graphic displays.'
> Petch and Haines-Young, 1986.

Because of its spatial basis, GIS will always retain some degree of separateness. There will always be a need for specialist firms to write spatial processing software. There will always be a niche for specialist geo-information vendors. There will be a continuing, hopefully, growing market for *spatial* information specialists within companies who know when, why and how to use GIS functions to produce helpful information products. On the other hand, we do foresee GIS becoming generally more integrated into the business computing mainstream and thus becoming more subject to the conventional disciplines of corporate Information Systems planning.

THE IMPORTANCE OF ISDMs IN CORPORATE GIS

> 'From the plethora of dead, dying and wounded GIS projects lamented in the literature, GIS managers must face facts: GIS project implementation is a high risk undertaking where complete failure is not uncommon and marginal positive impact is common place. When walking the high wire, project managers must balance the hard and soft aspects of GIS implementation — failure to fully appreciate organisational issues and employ appropriate effective project management techniques will inevitably result in a nasty fall.'
> Source: Diploma student.

One benefit of GIS becoming more integrated into the mainstream of corporate computing may be that disciplines and attitudes derived from mainstream ISDMs may more frequently be applied to GIS project planning and implementation. The major thrust of this book has been to argue that, in particular, socio-technical ISDM approaches, with their emphasis on reconciling technical opportunities with organisational realities, offer a rich source of ideas which might help GIS managers to fall less frequently from the high wire.

At this point, however, we should admit the reality that at present there is very little evidence of *any* large-scale adoption of formal socio-technical methodologies in GIS. In the UK, there have been attempts to promote the use of the SSADM (Structured Systems Analysis and Design Methodology) for use in major GIS projects. SSADM is a government sponsored methodology and the CCTA (1994) have produced a booklet describing GIS extensions to SSADM. As a methodology, however, SSADM based upon the technical aspects of process and dataflow analysis rather than deriving from the socio-technical roots of methodologies such as ETHICS or Multiview. Also, although SSADM has undoubtedly been used in some major GIS projects, it does not appear to have been generally adopted. In a recent survey of 21 major Irish GIS projects, O'Donaghue (1997) reports that he found that only 6 had adopted a formal methodology, 3 using SSADM and 3 traditional Systems Analysis: 'One of the most surprising facts to emerge from the analysis... is the small number of organisations which used any recognised ISDM at all'.

Overwhelmingly, the impression one gets from reading published accounts and listening to the sometimes rueful accounts of GIS projects given by practitioners is of a pragmatic, commonsense, 'buy the kit and things are bound to fall into place' sort of approach being adopted. People seem genuinely surprised when their projects are buffeted by changing political climates within management; when vendors fail to deliver promised functionality or delivery dates; when colleagues don't immediately become converts when they are shown the technological marvels which are being offered to them. There has seemed often to be an assumption that, generally speaking, in most cases things will work out right. Wrong! In many cases IT projects, and indeed many GIS projects, do not work out as expected. Projects fail, and even eventually successful projects have their sticky moments. Rather than assume success, our argument is that it would be better to acknowledge the full complexities of persuading organisations to adopt a technology which might radically alter existing patterns of behaviour and thus to employ an appropriate ISDM methodology to help guide the adoption process.

With regard to what constitutes an 'appropriate' ISDM for GIS projects, earlier chapters have made clear our view that broadly socio-technical and RAD approaches have much to recommend them. As ISDMs have developed in mainstream computing the *frontier* of their concerns have expanded to encompass ever-broader areas of organisational behaviour. Initially methodologies drew a fairly tightly defined boundary around their area of concern, their focus being primarily upon the *technology* of delivering information systems. Hoping to improve the success rate of IS projects, however, the authors of ISDMs have progressively broadened the scopes of their models in order to embed information systems development within an understanding of the broader business contexts of host organisations. The objective of methodologies such as ETHICS, Multiview, Organic Life Cycle, RAD and Evolutionary Delivery as outlined in Chapter Four is quickly to deliver information systems which are appropriate to the business needs of host organisations. It seems to us that this is exactly what GIS champions should focus on and, therefore, that such methodologies have much to offer to GIS project planning.

Furthermore, it seems to us that the development of Intranets, Extranets and 'virtual organisations' will soon further emphasise the need for a socio-technical approach to information systems development. In earlier chapters we have considered the difficulties of reconciling alternative viewpoints within companies. Often GIS projects have become footballs kicked around competing departments and power groups. Soon, if not already, however, the frontier of GIS and of ISDMs will shift from developing systems *within* organisations to developing them *between* organisations. This will add a still further layer of complexity and ISDMs will clearly need to develop ways of encompassing the politics of developing major projects involving many powerful organisations.

CONTINGENCY AND THE REALITY OF 'MUDDLING THROUGH'

Although we have emphasised the need for a greater awareness of socio-technical methodologies in GIS, it is important to recognise that it is possible to push this argument too far. ISDMs are 'models', that is they are idealised, generalised statements of how projects should be developed. In Chapter Six, however, we emphasised how varying circumstances and cultures within organisations can alter the chances of a project's success. What is appropriate in one company may not be in another. Even within one organisation a single ISDM will not be universally appropriate. The idea of *the* corporate methodology doesn't work. Methodologies and methods are appropriate only to a level in the organisation, to the role or business position of the GIS (or IS) and to the stage in a corporate development life cycle of IS. And even then no methodology is a perfect fit.

Any methodology, unthinkingly applied, will produce tensions because its assumptions, which are fixed, do not square up with the reality of business and cultural situations, which are complex, elusive and transitory. So we are forced, in a mixture of frustration and expectancy to recognise the *contingency* of project development. Ultimately, all circumstances are unique and all methods in practice are equivocal. The application of methodologies is dominated by the need to recognise contingencies so that specific local and temporary circumstances can be dealt with. No model should be slavishly followed, only adapted to circumstance. No methodology can be completely prescriptive, only a guideline. There is no masterplan that will always work.

In the GIS field the reality of development situations both for users and development expert is often best described by the phrase 'muddling through'. Things get done despite the methodology. The question we must ask then is, 'what use is our knowledge of development methodologies?' The answer is a paraphrase of the answer to the general question about the use of knowledge. It allows us to understand those things we have little or no experience of and to put our empirical knowledge into a framework. Knowing that there are issues of interface design or of human activity systems and so on gives us a repertoire of ideas against which we can codify our own directly acquired knowledge. This gives us versatility in complex situations and an understanding of risk that we would not otherwise have. So when muddling through we can with more confidence implement ad hoc measures and take action upon limited data. We can know of and share goals and visions of systems without a full and structured articulation of a plan. Muddling through is always going to happen. An understanding of ISDMs allows systems developers to muddle through with style!

The advantage of being aware of the wide variety of ISDMs which are available from mainstream computing is that such wide awareness increases the flexibility of response which project leaders can bring to their projects. A project leader who clings unthinkingly to a single methodology is very much a 'one-club golfer': No matter that the problem hit it with a nine-iron.

Sadly, there is little evidence of widespread knowledge of alternative ISDMs being commonplace within the GIS community at present. In this respect, Waters (1993) recounts a pertinent experience. He says he tried to interest two groups of graduate GIS students in the differing philosophies that underpin alternative ISDMs. With one group the response, seemingly, was one of a blank inability to see the relevance that such ideas might have for GIS. The other group saw the light but wondered why it had taken their superiors so long to catch on:

> 'I have submitted these ideas in rudimentary form to graduate students studying GIS... In doing so I was presented with two difficult questions. [One] group essentially asked, "What's the difference? Will these paradigms really lead to new GIS implementations?" The short answer is, "Yes, it is different. Just look at the computer science literature." The [other] group posed the question, "What has taken the GIS community so long to recognise these alternative approaches?" The only answer to that is "What indeed!" '
> Waters, 1993.

We hope that, increasingly, the blank incomprehension response will diminish and that the ideas touched upon in this book will become a common currency within the GIS community. The more that GIS is understood to be a *socio*-technical enterprise, the greater the likelihood of success.

Glossary

Attribute Data: Data which describes the condition of a spatial object, for example, the crop which is being grown in a field or the age of a building.

Benchmarking: Assessing the performance of software and hardware against a series of predetermined tests.

Big-bang Implementation: Switch off the old system one day. Switch on the new system the following day. A brave and high profile way of going live with a system.

'Blue Print' Planning: A method of planning which assumes that there is a fixed goal towards which a project can be stirred.

Boston Matrix: A simple cross classification matrix used to assess the business prospects of products and projects.

Business Process Outsourcing: Sub-contracting a previously internal process to another organisation. The development of networks, and particularly extranets, is encouraging outsourcing of information-based processes.

Business Process Reengineering (BPR): Radical re-structuring of tasks, workflows, lines of responsibility and modes of operation consequent upon the introduction of New Technology.

Business Strategy: A statement of an organisation's fundamental goals and how they are to be realised.

CASE Tools: Computer Aided Software Engineering Tools. These are high-level packages which produce code, often through visual interfaces, such as in database design, where EAR Diagrams can be produced graphically. They often involve the construction of object libraries of re-usable code that can be built up using the high level tools into complex data processing modules.

CATWOE: The main elements of soft systems analysis, clients (C), actors (A), the transformation (T), the Weltanschauung (W), ownership (O) and the environmental constraints (E) of the system.

Client Complexity (selection grids): The degree of harmony of objectives between clients for a project — unitary, pluralist or coercive.

Client/Server: Networks in which PCs (clients) draw files and support services from a more powerful/larger disk capacity (file) server PC.

Code-and-Fix: An approach to software development devoid of methodology.

Componentware: A term used to describe the modularisation of software products in a common operating environment which can be linked together.

Computer-mediated Work: Work which is done through the intermediary of a computer, i.e. GIS operatives work on electronic representations of maps, not maps themselves.

Conceptual Model: In soft systems analysis, a logical description of information processes.

Corporate Strategies: The high level policies by which organisations intend to implement their objectives.

Cost avoidances: Costs which were incurred by a manual system which will not be incurred in a computerised system because a process is no longer necessary.

Cost–Benefit Analysis (CBA): Formal estimation of the ratio of a project's benefits against its costs expressed in monetary terms.

Critical Success Factors (CSFs): A key factor which is seen as being essential to achieving a business objective.

Data Warehouse: A central repository database into which data from operational databases are copied. Within the data warehouse, data are held in formats which make them amenable to analytical processing. The purpose of data warehousing is to build a picture of an organisation's activities which provides policy relevant information.

Decision Support Prototyping: One type of essential prototyping concerned with subjecting a system to tasks at different levels of management in order to refine data structures and operations between levels.

Decision Support Systems (DSS): Interactive Information Systems which provide decision makers with diagrams, tables, forecasts, etc., to help managers to take decisions.

Demand Pull: Occurs when information systems are introduced into organisations through business demands (see Technology Push).

Design Prototyping: Prototyping to change or develop systems specifications or performance.

Deskilling: The removal from a job of skills consequent upon the introduction of a computer system.

Disintermediation: The process of 'removing the middleman' which Information Technology is allowing to take place, leading to flatter organisational structures.

Disjointed Incrementalism: An approach to project implementation which recommends that, rather than having a fixed, blueprint plan from which to work, project managers should be opportunistic and flexible as they proceed incrementally towards their goal.

Electronic Data Interchange (EDI): A standard format for exchanging business data. An early form of electronic commerce.

Empowerment: The increased ability of workers to control their working environment consequent upon the introduction of computer systems. Increased worker autonomy.

Encapsulated: A term used in Object-Oriented programming to describe the integration of data and functions within objects.

End-state Planning: Similar to 'Blue Print' Planning.

Enterprise Information System: An Information System, which is pervasive throughout an organisation. Ideally all the organisation's data is held within one system and access to this system is available to all relevant employees.

Entitation: The process of recognising and defining entities.

Essential Prototyping: Prototyping to derive systems requirements.

ETHICS: Effective Technical and Human Implementation of Computer Systems. A methodology for systems analysis and design based on socio-technical principles.

Event Prototyping: One type of Essential Prototyping concerned with testing responses to events.

Everyone's Information System: A trendy way of referring to an Enterprise Information System.

Evolutionary Delivery: An IS development methodology based on the idea of modularised, incremental, adaptive stages of delivery.

Executive Information Systems (EIS): Interactive, easy to use, systems intended to provide executives with pertinent summaries of key information.

Expert Imposition: The delivery of Information Systems which the 'experts' believe workers need, rather than which workers themselves have helped to design.

Extranet: The use of Web technologies to allow suppliers or customers to access previously internal flows of information, thus blurring the boundaries of the organisation.

Fetishism of the Product: A Marxist term which expresses the tendency of consumers to covet products as objects in their own right, rather than for their use values.

Functional Decomposition: The progressive breaking down of a system into smaller, and more tractable, sub-systems.

GIS Champion: An employee who is committed to introducing GIS into an organisation.

GIS Godfather: An executive-level employee who is prepared to assist the introduction of GIS into an organisation.

Human Activity System: Any system which involves people, such as a business.

Human Computer Interface (HCI): The means by which a person interacts with a computer, the screen, mouse, keyboard, sound system, etc.

Information Resource Management (IRM): The conscious management of an organisation's information, in recognition that 'information' is a key corporate resource.

Information Strategy: A statement identifying the information services, which an organisation needs to service its business strategy, and indicating how these services will be provided.

Information Systems Development Methodology (ISDM): A sequence of stages, techniques, and tools, which are intended to help systems analysts to develop successful information systems. Based of an underlying philosophy of how organisations work.

Inheritance: The property of computer objects such that they take on the attributes and methods of the hierarchical class of which they are a member.

Interoperability: The ability of computer software, produced by different vendors, to work seamlessly together. Interoperability is a major current goal for the computer industry generally and is being strongly pursued within the GIS industry.

Intranets: The use of Internet tools (HTML, browsers, search engines etc.) to establish private 'internets' within companies. The advantages of Intranets over client/server networks are: cheaper tools (most internet tools are free), lower skill requirements, easier delivery throughout the organisation.

Joint Applications Development (JAD): The practice of getting the users or customers for systems involved in specifying what the system should do, how it should perform and what it should be like to use. JAD sessions are, in effect, a means of prototyping.

Management Information System (MIS): An information system which provides managers with reports about the performance of the operational areas of an organisation.

Messages: The mechanisms by which computer objects communicate and respond to each other.

Milestone: An application or product which is delivered before the completion of the entire project to allow the organisation to gain interim benefits and to maintain corporate support.

Mission critical Systems: Computer systems upon which the survival of an organisation depends.

Mission Statement: A formal statement of the fundamental long-term aims of an organisation.

Multiview: A methodology for systems analysis and design based on taking a multi-viewed approach including soft systems and technical systems approaches.

Network Computers: Cheap, cut down (no disk storage, minimal operating system) PCs which will draw not only data but also applications (Applets) from networks.

Object: In database technology the term refers to the structuring of information such that attributes of things (objects) are encapsulated with rules for information processing and with relations to other objects.

Objectives: Strategic targets which derive from an organisation's Mission Statement.

Object-Orientation (O-O): Analysing, designing and operating an information system using the concepts and/or technology of objects.

Online Analytical Processing (OLAP): The use of analytical tools (statistics, cluster analysis, neural nets, mapping tools etc.) to search for patterns in data, without hindering the performance of mainstream transaction processing databases.

Open Systems: Systems that use standards to enable operation of software from different suppliers and share data in different native formats.

Organic Life Cycle: An ISDM based on the idea of prototyping and the spiral model of development.

Parallel Running: Running a new computer system alongside the manual old. Incurs costs but allows bugs to be ironed out of a new system before the manual system is dispensed with.

Participation: The term used in ETHICS to describe the formal structures for involvement of people in the analysis and design of information systems. It takes various forms — consultative, democratic, responsible.

Performance Indicators: Measurements which indicate whether Critical Success Factors and Objectives are being achieved.

Phased Run in: Introducing a new system gradually — a function or a department at a time.

Polymorphism: The property of objects in O-O systems of being able to respond to the same messages in different ways.

Power webs: Informal networks within organisations which can exert influence upon decision making.

Primitive Prototyping: One type of Essential Prototyping concerned with verifying parts of data-models.

Prototyping: The construction of a working model provided to help potential users to determine their requirements.

Rapid Application Development (RAD): A term applied to a collection of practices first defined by James Martin (1991) as consisting of JAD Sessions, prototyping, SWAT teams, timeboxed deliverables and CASE Tools. It is a collection of practices rather than a single methodology but the various practices can be structured in a methodology. The practices prove to be effective and versatile in a wide range of applications and to deliver rapid development in specific areas.

Relational Model: The dominant model for the design of database management systems during the 1980s, which allows users to believe that data are held in 'tables'.

Request for Information (RFI): Questionnaires sent to vendors, and other organisations, in order to gain information about the suitability of products for a project.

Reskilling: Retraining employees to provide them with the new skills necessary to be able to take advantage of new technologies (see Deskilling).

Re-usability: The characteristic of objects which relies on the fact that they can be treated as discrete components which can be used again and again.

Rich picture (Soft Systems Analysis): A stylised cartoon which expresses the analyst's understanding of a problem area.

Root Definition (Soft Systems Analysis): A root definition is a concise, tightly structured description of a human activity system which states what the system is.

(corporate) Sharks: Employees who exploit the political natures of organisations excessively.

Skunkworks: Teams of employees that are independent of normal departmental and reporting structures which are created temporarily to focus upon the delivery of a particular project. The objective is to achieve the levels of commitment, team spirit and flexibility which are conventionally associated with small firms within large organisations.

Socio-technical Computing: In general, a view of computing which encompasses the human and organisational aspects of computing as well as the technical aspects. Also the term used to describe a particular information system development methodology (The Socio-technical approach; Mumford, 1981).

Soft Systems Analysis: A problem structuring technique originated by Checkland (1981) which attempts to use concepts taken from conventional systems theory in an interpretative and humanistic manner rather than in a formal and technocratic one.

Spatial Decision Support Systems (SDSS): GIS systems which are designed to help managers to take spatially based decisions.

(corporate) Spiders: Employees who use their networking skill to develop influence within their organisations in excess of their formal positions.

Spiral Model: A life-cycle model which has multiple sequences of the problem–analysis–design–test–implement cycle.

Stakeholder: Any person, or group of persons, who will be affected by a project.

Steady-state (of a system): The configuration of an open system which is in dynamic equilibrium.

Structured Systems Analysis and Design Methodology (SSADM): A set of methods developed and promoted by the UK Central Communications and Telecommunications Agency for the analysis and design of information systems. The methods are in the public domain.

SWAT Teams: The term is borrowed from the police where it means Strategic Weapons Assault Team. In software engineering it means 'skilled with advanced tools'. It is an acronym for groups of specialist software developers who are brought into projects to focus on specific areas of development for rapid delivery usually using advanced tools.

SWOT Analysis: A tool for strategic thinking which requires managers to identify **S**trengths, **W**eaknesses, **O**pportunities, and **T**hreats to their organisations.

System Complexity (Selection Grids): The level of the technical challenge involved in an IS project.

Task Scope: The proportion of an entire process which lies within a single job description.

Task Variety: The number of activities within a job specification.

Technical Prototyping: A type of design prototyping concerned with testing the functionality and performance of software.

Technical Subsystem: That part of an information system which is the technology, viz. hardware and software.

Techniques (ISDMs): A method of manipulation or analysis used to provide a result relevant to a stage within an Information Systems Development Methodology.

Technological Imperative: The belief that technology once discovered will inevitably be used. Technology creates its own demand.

Technology Push: Occurs when information systems are introduced into organisations because the technology is available and fashionable (see Demand Pull).

Tools (ISDMs): A piece of software designed to support a technique within an Information Systems Development methodology.

Transaction Processing System (TPS): A 'transaction' is an event which generates or modifies data in an information system. A transaction processing system is an information system set up to handle business transactions.

User-interface Prototyping: A type of design prototyping concerned with testing the effectiveness and efficiency of the use of systems by people.

User Needs Study (UNS): Research conducted within an organisation in order to determine the demand which there may be for a new technology or system. Often based on interviews and questionnaires.

Value Chain: The sequence of transformations which create a product's value for customers.

Virtual Organisations: Organisations which, in reality, exist only on the Internet. Customers may believe that they are dealing with a conventional organisation which has a single physical reality but in reality the constituent parts of the 'organisation' may be scattered widely and may be using the Internet to co-ordinate their activities.

Weltanschauung: Literally 'world view'. The perspective from which someone views a problem.

Workbench (ISDMs): An integrated set of software tools which support an Information Systems Development Methodology through all or some of its stages.

References

ABLER, R., 1987, The National Science Foundation Center for GIS. *International Journal of Geographical Information Systems,* **1**, 4, pp. 303–326.

ADAMS, S., 1996, *The Dilbert Principle: A Cubicle's Eye View of Bosses, Meetings, Management Fads and other Workplace Afflictions,* (London: Boxtree).

AHMED, G., 1993, A warning about programmers. *Computing,* 7/1/93, p. 8.

ALLEN, C. P., 1991, *Effective Structured Techniques from Strategy to CASE,* (London: Prentice-Hall).

ALTER, S., 1992, *Information Systems: A Management Perspective,* (New York: Addison-Wesley).

ANGELL, I., 1993, I don't give pat answers. *Computing,* 7/1/93, p.10.

AUTHES, G. H., 1993, Phoenix maps out GIS plan. *Computerworld,* January, p.82.

AVISON, D. E. and FITZGERALD, G., 1988, *Information Systems Development: Methodologies, Techniques and Tools,* (Oxford: Blackwell).

AYBET, J., 1996, The role of GIS in Business Process Reengineering. *Association for Geographic Information Conference Proceedings,* 2.7.1–2.7.4, (London: AGI).

BARKER, C., 1998, London Ambulance Service gets IT right. *Computing,* 25/6/98, p. 20.

BARR, R., 1991, A federal approach to GIS. *Mapping Awareness,* **5**, 6, pp. 15–19.

BECKER, P., CALKINS, H., COTE, C., FINNERAN, C., HAYES, G. and MURDOCK, T., 1997, *GIS Development Guide,* (Albany, New York: Local Government Technology Services, State Archives and Records Administration), (http://www.ncgia.ucsb.edu/education/curricula/giscc/units/u136/).

BELL, S. and WOOD-HARPER, T., 1992, *Rapid Information Systems Development: A Non-Specialist's Guide to Analysis and Design in an Imperfect World,* (London: McGraw-Hill Book Co.).

BERRY, J. K., 1996, Don't forget the human factor in GIS. *GIS World,* **9**, 7,pp. 28–29.

BESTEBREURTJE, J. G. A., 1997, GIS Project Management. Unpublished MSc thesis, Department of Environmental and Geographical Sciences, Manchester Metropolitan University.

BIRKEN, M., CLARKE, G., CLARKE, M., and WILSON, A., 1996, *Intelligent GIS: Location Decisions and Strategic Planning,* (Cambridge: GeoInformation International).

BLENHEIM ONLINE, 1993, What do you mean GIS?, *Preview GIS '93,* (London: Blenheim Online).

BLOCK, R., 1983, *The Politics of Projects,* (New York: Yourdon Press).

BOEHM, B. W., 1981, *Software Engineering Economics,* (Englewood Cliffs, NJ: Prentice-Hall).

BOEHM, B. W., 1988, A Spiral Model of software development and enhancement. *Computer,* May, pp. 61–72.

BRADBURY, D., 1995, Planning for a change. *PC Week,* 7/3/97, pp. 8–9.

BROMLEY, R. D. F. and COULSON, M. G., 1989, *Geographical Information System and the Work of a Local Authority: The Case Study of Swansea City Council,* (Swansea: Department of Geography, University College of Swansea).

BUCHANAN, D. A., 1993, The Organisational Politics of Technological Change. In Medyckyj-Scott, D. and Hearnshaw, H. M., (Eds.), *Human Factors in Geographical Information Systems,* (London: Belhaven Press).

BUSINESS COMPUTER WORLD, 1997, Where in the world. *Business Computer Weekly,* January, p. 18.

BUTTON, K., 1996, Thou shalt not. *Computer Weekly,* 2/5/96, pp. 32–33.

CAMBRIDGE COMPUTER CONSULTANTS & DAN RICKMAN ASSOCIATES, 1993, *GIS in Government: Realizing the Opportunities, CCTA,* (London: HMSO).

CAMPBELL, H. and MASSER, I., 1995, *GIS and Organisations: How Effective are GIS in Practice?,* (London: Taylor and Francis).

CCTA, 1994, *SSADM for Handling Geographic Information,* (London: HMSO).

CCTA, 1995, *Object Orientation and SSADM,* (London: HMSO).

CHECKLAND, P., 1981, *Systems Thinking: Systems Practice,* (Chichester: John Wiley).

CHECKLAND, P. and SCHOLES, J., 1990, *Soft Systems Methodology in Action,* (Chichester: John Wiley).

CHRISMAN, N., 1987, The design of Geographical Information Systems based on social and cultural goals. *Photogrammetic Engineering,* **53**, 10, pp.1367–1370.

CLARKE, G. and CLARKE M., 1995, The development and benefits of customized spatial decision support systems. In Longley, P. and Clarke, G., *GIS for Business and Services Planning,* (Cambridge: GeoInformation International).

CONSTANT, E. W., 1994, The social locus of technological practice: community, system or organisation? In Bijker W. E., Hughes T. P. and Pinch T. J. (Eds.) *The Social Constructions of Technological Systems: New Directions in the Sociology and History of Technology,* (Cambridge, MA: MIT Press).

CONSTANTINE, L. L., 1995, *Peopleware,* (Englewood Cliffs, N.J: Yourdon Press).

COOPERS and LYBRAND, 1988, Company Profile: Coopers and Lybrand. *Mapping Awareness,* **2**, 5, pp. 42–45.

COOTE, A., 1997, Running a successful GIS project — a supplier's perspective. *Association for Geographic Information Conference Proceedings,* 9.1.1–9.1.4, (London: AGI).

CROSWELL, P. L., 1989, Facing reality in GIS implementation: lessons learned and obstacles to be overcome. *URISA Conference Proceedings,* **4**, pp. 15–35.

CULLIS, B. J., 1994, A strategy for assessing organizational GIS adoption success. *GIS/LIS 94,* pp. 208–217.

DEMARCO, T. and LISTER, T., 1987, *Peopleware: Productive Projects and Teams,* (New York: Dorset House).

DEARNLEY, P. A. and MAYHEW, P. J., 1983, In favour of systems prototypes and their integration into the systems development cycle. *The Computer Journal,* **26**, pp. 36–42.

DEMERS, M. N., 1996, *Fundamentals of Geographic Information Systems*, (New York: John Wiley and Sons Inc.)

DENSHAM, P. J., 1991, Spatial Decision Support Systems. In Maguire, D. J., Goodchild, M. F. and Rhind, D. W., (Eds.) *Geographical Information Systems Principles and Applications,* (London: Longman).

DICKINSON, H. J. and CALKINS, H.W., 1988, The economic evaluation of implementing a GIS. *IJGIS,* **2**, 4, pp. 307–329.

DICKINSON, H. J. and CALKINS, H.W., 1990, Comments on 'Concerning the economic evaluation of implementing a GIS'. *IJGIS,* **4**, 2, pp. 211–213.

DEPARTMENT of the ENVIRONMENT, 1987, *Handling Geographic Information: Report of the Inquiry Chaired by Lord Chorley,* (London: HMSO).

DOWNS, A., 1967, A realistic look at the final payoffs from urban data systems. *Public Administrative Review,* **27**, pp. 204–10.

DUC, R. T. and HENDERSON-SELLERS, B., 1995, The changing paradigm for object project management. *Object Magazine,* July–August, pp. 55–60 and p.78.

DUNN, R. and HARRISON, A., 1992, Assessing user requirements for GIS — the critical stage in implementation: how prototyping can help you get it right. *Mapping Awareness Conference Proceedings.*

EASON, K., 1988, *Information Technology and Organisational Change,* (London: Taylor and Francis).

EASON, K., 1993, Planning for change: introducing a GIS. In Medyckyj-Scott, D. and Hearnshaw, H. M., (Eds.), *Human Factors in Geographical Information Systems,* (London: Belhaven Press).

EDWARDS, C., WARD, J. and BYTHEWAY, A., 1991, *The Essence of Information Systems*, (London: Prentice Hall).

ENGLAND, J., 1996, The road to a corporate GIS in Gloucestershire. *Mapping Awareness,* **10**, 4, pp. 20–23.

FERGUSON, W., 1990, Unit 61, Functional Requirements Study. In Goodchild, M. F. and Kemp, K.K., (Eds.) 1990, *NCGIA Core Curriculum in GIS,* National Center for Geographic Information and Analysis, (Santa Barbara: University of California).

FERRARI, R. and ONSRUD, H. J., 1995, *Understanding Guidance on GIS Implementation: A Comprehensive Review.* Technical Report 95–13, NCGIA, (Oraono: Department of Spatial Science and Engineering, University of Maine).

FLYNN, D. J., 1992, *Information System Requirements: Determination and Analysis,* (London: McGraw-Hill).

FOTHERINGHAM, A. S., 1996, Foreword. In *GIS Innovations '95.* (London: Taylor and Francis).

GIDDINGS, R. V., 1984, Accommodating uncertainty into software design, *Communications of the Association for Computer Machinery,* **27**, pp. 428–434.

GILB, T., 1988, *Principles of Software Engineering Management,* (Wokingham: Addison-Wesley).

GILL, S., 1994, Introducing GIS into Powys. Unpublished MSc Thesis, UNIGIS programme.

GLOVER, J., 1996, The integration and interoperability challenge — the need for OpenGIS. *Association for Geographic Information Conference Proceedings,* 2.20.1–2.20.5 (London: AGI).

GOODCHILD, M. F. and KEMP, K. K., (Eds.) 1990, *NCGIA Core Curriculum in GIS,* National Center for Geographic Information and Analysis, (Santa Barbara: University of California).

GOODCHILD, M. F. and RIZZO, B. R., 1987, Performance evaluation and workload estimation for Geographic Information Systems. *International Journal of Geographical Information Systems,* **1**, pp. 76–77.

GOODWIN, C., 1995, Left on the bench, *Computing,* 25/9/95, 8.

GOSSETTE, F., FERGUSON, W. and DUEKER, K., 1990, Unit 60, System Planning Overview. In Goodchild, M. F. and Kemp, K. K., (Eds.) 1990, *NCGIA Core Curriculum in GIS,* National Center for Geographic Information and Analysis, (Santa Barbara: University of California).

GRIMSHAW, D. J., 1994, *Bringing Geographical Information Systems into Business,* (Harlow: Longman Scientific and Technical).

GRIMSHAW, D. J., 1996, Will the virtual organisation need a real GIS?, *Association for Geographic Information Conference Proceedings,* 2.6.1–2.6.5, (London: AGI).

GRIMSHAW, D. J. and SCHOLTEN H., 1997, A comparison of the implementation of GIS in the motor vehicle industry, *Joint European Conference on Geographical Information '97 Proceedings,* **2**, pp. 1187–1196.

GRINDLEY, K., 1993, *Managing I.T. at Board Level: The Hidden Agenda Exposed,* (London: Pitman).

GUNTHER, O., 1997, From GISystems to GIServices: spatial computing on the Internet marketplace, *Interop '97 Conference Proceedings,* 78–80, (Santa Barbara: NCGIA).

HAMMER, M. and CHAMPY, J., 1993, *Reengineering the Corporation: A Manifesto for Business Revolution,* (London: Nicholas Brearley).

HANDY C., 1989, *The Age of Unreason,* (London: Hutchinson).

HARTLEY, J. L., HOMER, I. R., TROWS, S., HINTON, P. H. and EVERT, C. 1992, Inter utility exchange of electronic map based records in the north east of England. *Association for Geographic Information Conference Proceedings,* 1.191–1.19.6, (London: AGI).

HIRSCHEIM, R. and KLEIN, H. K., 1992, Paradigmatic influences on Information Systems. *Advances in computers,* **34**, pp. 293–391.

HOBSON, S. A., 1991, Methodology issues in GIS Introduction. *AM/FM Conference Proceedings,* pp. 207–211.

HUXHOLD, W.E., 1991, *An Introduction to Urban Geographic Information Systems,* (New York: Oxford University Press). (Chapter 7: The Model Urban GIS Project).

HUXHOLD, W. E. and LEVINSOHN, A. G., 1995, *Managing GIS Projects,* (New York: Oxford University Press).

INNES, J. E. and SIMPSON, D., 1993, Implementing GIS for planning. *American Planning Association Journal,* **59**, 2, pp. 230–236.

IVES, M. J. and CRAWLEY, K. J., 1995, GIS implementation issues. In Green, D. R. and Rix, D., *AGI Source Book for Geographic Information Systems 1995,* (London: AGI).

JAYARATNA, N., 1994, *Understanding and Evaluating Methodologies.* (London: McGraw Hill).

KEEN, P., 1981, Information Systems and Organisational Change. *Communications of the Association for Computer Machinery,* **24**, 1, pp. 24–33.

KEENAN, M., 1996, Structured Design Methodologies for GIS. Unpublished MSc thesis, UNIGIS programme, Manchester Metropolitan University.

KENDRICK, G. and BATTY, P., no date, Smallworld GIS: Use of an integrated CASE tool for GIS customisation, *Smallworld Technical Paper,* 11, (Cambridge: Smallworld).

KORTE G., 1996, Weighing GIS benefits with financial analysis. *GIS World,* **9**, 7,48–52.

KUIPER, D., 1992, *Those Idiots in the Computer Room: Computer Myth-conceptions Spelled Out in Plain English,* (Portland: MacAdam House Publishing).

LAWRENCE, V. and PARSONS, E., 1997, GIS in disguise: improving decision-making with geography. *Association for Geographic Information Conference Proceedings,* 3.2.1–3.2.5, (London: AGI).

LAY, P., 1985, Beware of the Cost Benefit model for IS project evaluation. *Journal of Systems Management,* pp. 30-35.

LEGG, S., 1997, Benefits of RAD methodology for GIS projects. *Association for Geographic Information Conference Proceedings,* 12.2.1–2.2.4, (London: AGI).

LEVINSOHN, A. G., 1997, Enterprise GIS gains prominence in Canada. *GIS World,* **10**, 4, pp. 60–62.

LEYDEN, J., 1998, GPS charts 999 status. *Network News,* 24/7/98, 22.

LITECKY, A., 1981, Intangibles in Cost/Benefit Analysis. *Journal of Systems Management,* pp. 15–17.

LOCAL GOVERNMENT MANAGEMENT BOARD, 1992, *GIS: Functional Specification Version 1,* (Luton: LGMB).

LOCAL GOVERNMENT MANAGEMENT BOARD, 1993, *Benchmarking Guidelines: A Guide to Evaluation,* (Luton: LGMB).

LOCAL GOVERNMENT MANAGEMENT BOARD, 1995, *GIS: Go with the Flow,* (Luton: LGMB).

MAHONEY, R. P., VALE, M. J., and ALVES, R., 1997, The development of a rigorous Cost/Benefit Case. *Joint European Conference on Geographic Information Proceedings,* **2**, pp. 1177–1186, (Amsterdam: IOS Press).

MARBLE, D., 1995, An introduction to the structured design of Geographical Information Systems. In Green, D. R. and Rix, D., *Association for Geographic Information Source Book for GIS 1995,* (London: AGI).

MARTIN, D., 1993, Council invites experts to rethink IT Plan. *Computing,* 3/6/93, 3.

MARTIN, J., 1991, *Rapid Applications Development,* (London: MacMillan).

MARTIN, J., 1997, *Cybercorp: The New Business Revolution,* (London: McGraw-Hill).

MASSEY, J., 1997, University Challenge. *Computing,* 20/3/97, 46.

MCCONNELL. S., 1996, *Rapid Development: Taming Wild Software Schedules,* (Redmond: Microsoft Press).

MCFARLAN, F. W., 1981, Portfolio approach to Information Systems. *Harvard Business Review,* pp. 142–150.

MCLAREN, R. A., 1992, Implementing a GIS: A review of alternative strategies. *Mapping Awareness Conference Proceedings,* (London: Blenheim Online), pp. 17–32.

MEDYCKYJ-SCOTT, D. J. and CORNELIUS, S., 1993, User viewpoint: a survey. In *Geographic Information Systems Report,* (Uxbridge: Unicorn Seminars Ltd.).

MOWSHOWITZ, A., 1976, *The Conquest of Will: Information Processing in Human Affairs,* (Reading: Addison-Wesley).

MUMFORD, E., 1983, *Designing Human Systems — The ETHICS Method*, (Manchester: Manchester Business School).

MUMFORD, E., 1995, *Effective Systems Design and Requirements Analysis: The ETHICS Approach*, (London: MacMillan).

MUSGRAVE, T., 1966, Ignore GIS — information management is the key. *Association for Geographic Information Conference Proceedings*, 2.8.1–2.8.6, (London: AGI).

NAHAPIET, J. E., 1984, Assessing costs and benefits in system design and selection. In Otway, H. J. and Peltu, M., *The Managerial Challenge of New Office Technology*, (London: Butterworth).

NAUGHTON, J., 1984, *Soft Systems Analysis: An Introductory Guide*, (Milton Keynes: Open University Press).

NEWTON, P. W., ZWART, P. R. and CAVILL, M. E., 1995, *Networking Spatial Information Systems*, (Chichester: John Wiley).

O'DONAGHUE, D., 1997, Matching Methodology with Structure. Unpublished MSc Thesis, UNIGIS programme.

OBERMEYER, N.J. and PINTO, J.K., 1994, *Managing Geographic Information Systems*, (New York: Guilford Press).

OFFEN, G., BROOKE, A. and TIMMS, T., 1997, Reengineering business processes at The Coal Authority with fully integrated GIS and workflow. *Association for Geographic Information Conference Proceedings*, (London: AGI).

OPENSHAW, S., CROSS, A., BRUNSDON, M. and LILLIE, J., 1990, Lessons learned from a failed GIS. *Association for Geographic Information Conference Proceedings*, 2.3.1–2.3.5, (London: AGI).

PARKER, M. M., BENSON, R. J. with TRAINOR, H. E., 1988, *Information Economics*, (Englewood Cliffs, NJ: Prentice-Hall).

PARSONS, T. and SHILS, E., 1951, *Towards a General Theory of Action*, (Harvard, Mass.: Harvard University Press).

PELTU, M., 1996, Death to cuts, *Computing*, 9/5/96, 34.

PETCH, J. R and HAINES-YOUNG, R.H., 1991, *Physical Geography: Its Nature and Methods*, (London: Harper and Row).

PETERS, T., 1987, *Thriving on Chaos: Handbook for a Management Revolution*, (New York: Alfred A Knopf).

PEUQUET, D. J. and BACASTOW, T., 1991, Organisational issues in the development of GIS: a case study of US Army topographic information automation. *International Journal of Geographical Information Systems*, **5**, 3, pp. 303–319.

PINTO, J. K. and AZAD BIJAN, 1994, The role of organisational politics in GIS implementation. *URISA Journal*, **6**, 2, pp. 35–61.

PLANNING, 1993, Information for the Environment. *Planning Newsletter*, July.

PORTER, M. and MILLAR, V.E., 1985, How information gives you competitive advantage. *Harvard Business Review*, **63**, 4, pp.149–60.

REEVE, D. E. and WHEELER, R., 1991, Geographical Information Systems and Local Government Policy: The Kirklees Policy Mapping Project. *Local Government Policy Making*, **18**, 3, pp. 41–49.

REID, E., 1992, Machiavelli and data. *Association for Geographic Information Conference Proceedings*, (London: AGI).

ROBEY, D., 1987, Implementation and organisational impacts of information systems. *Interfaces*, **17**, pp. 72–84.

ROBSON, W., 1994, *Strategic Management and Information Systems: An Integrated Approach,* (London: Pitman).

ROYAL TOWN PLANNING INSTITUTE, 1992, *Geographic Information Systems: A Planner's Introductory Guide,* (London: RTPI).

SCHOFIELD, R., 1997, The Management of risk in a Large Data Capture Project. Unpublished MSc Thesis, UNIGIS programme.

SHERWOOD, N., 1995, Business geographics — a US perspective. In Longley, P. and Clarke, G., *GIS for Business and Service Planning,* (Cambridge: GeoInformation International).

SILK, D. J., 1991, *Planning IT: Creating an Information Management Strategy,* (Oxford: Butterworth-Heinemann).

SMITH, D. A. and TOMLINSON, R. F., 1992, Assessing costs and benefits of geographical information systems: methodological and implementation issues. *International Journal of Geographical Information Systems,* **6**, 3, pp. 247–56.

SMITH, B. and GOODWIN, R., 1996, The National Land Registry: feasibility study interim results. *Association For Geographic Information Conference Proceedings,* 4.4.1–4.4.2, (London: AGI).

SOMERS, R., 1994, GIS organisation and staffing, *URISA,* pp. 41–52.

SOMERS, R., 1994, Alternative development strategies. *GIS/LIS '94 Conference Proceedings,* (Phoenix),706–715.

SOMORJAY, M. and YEOMAN, B., 1996, National Land Information Service issues: confidentiality, liability, data: a local authority perspective. *Association For Geographic Information Conference Proceedings,* 4.4.1– 4.4.2, (London: AGI).

THURMAN, I., 1996, How Mace is using Geographic Information to define targeted store offers. *Association For Geographic Information Conference Proceedings,* 4.4.1– 4.4.2, (London: AGI).

TUTOR, D.J. and TUTOR, I.J., 1995, *System analysis and design: a comparison of structured methods,* (Oxford: NCC Blackwell).

WAINWRIGHT, P., 1996, Go with the flow. *Computing, 25/7/96,* pp. 22–25.

WARD, J., GRIFFITHS, P. and WHITMORE, P., 1990, *Strategic Planning for Organisations,* (Chichester: Wiley).

WATERS, N., 1993, GIS: paradigms lost. *GIS World,* **6**, 8, p.64.

WATERS, R., 1992, Kill the G in GIS! *Association for Geographic Information Conference Proceedings,* 2.5.1–2.5.3, (London: AGI).

WILCOX, D. L., 1990, Concerning the economic evaluation of implementing a CIS, *International Journal of Geographical Information Systems,* **4**, 2, pp. 203–210.

WILEY, L., 1997, Think evolution, not revolution for effective GIS implementation. *GIS World,* **10**, 4, pp. 48–51.

WILSON, J. D., 1997, Technology partnerships spark the industry. *GIS World,* **10**, 4, pp. 36–42.

WINFIELD, I., 1991, *Organisations and Information Technology: Systems, Power and Design,* (Oxford:Blackwell Scientific Publications).

WORRALL, L., 1994, Justifying investment in GIS: a local government perspective, *International Journal of Geographical Information Systems,* **8**, 6, pp. 545–566.

YOURDON, E., 1989, *Modern Structured Analysis,* (Englewood Cliffs, NJ: Prentice-Hall).

Index